ChaseDream GMAT 备考系列丛书

GMAT™ Focus Edition

批判性推理

逻辑分类精讲 第3版

毕 出 王钰儿◎编著

机械工业出版社
CHINA MACHINE PRESS

GMAT 批判性推理题以非形式逻辑作为依托，希望考生正确识别一个论证的结构，继而准确找到当前论证结构的评估方式，最终选出正确答案。本书正是以批判性推理题所考查的非形式逻辑为基础编写的。全书共三章，内容涉及批判性推理的基础知识、演绎论证、假说论证、批判性推理常见四大题型，以及最后的综合练习。本书语言简练，帮助读者从了解批判性推理的含义开始，进而掌握这种题型所有的考法及变化，最后快速而准确地作答。作者在书中揭示了 GMAT 批判性推理考题真正的考查重点，避免了市场同类书基于现象总结的弊端，具有强大的可推广性和适应性，内容专而不广，叙述简洁，通俗易懂。

本书适用于所有已经参加过或者准备参加 GMAT 考试的考生，也适用于喜好研究 GMAT 考试的同仁。

图书在版编目（CIP）数据

GMAT 批判性推理：逻辑分类精讲 / 毕出，王钰儿编著. —3 版. —北京：机械工业出版社，2024. 5
（ChaseDream GMAT 备考系列丛书）
ISBN 978 - 7 - 111 - 75851 - 8

Ⅰ. ①G… Ⅱ. ①毕… ②王… Ⅲ. ①逻辑推理 - 研究生 - 入学考试 - 自学参考资料 Ⅳ. ①O141

中国国家版本馆 CIP 数据核字（2024）第 100028 号

机械工业出版社（北京市百万庄大街 22 号 邮政编码 100037）
策划编辑：苏筛琴　　责任编辑：苏筛琴
责任校对：尹小云　　责任印制：单爱军
保定市中画美凯印刷有限公司印刷

2024 年 7 月第 3 版 · 第 1 次印刷
184mm × 260mm · 20. 25 印张 · 1 插页 · 501 千字
标准书号：ISBN 978 - 7 - 111 - 75851 - 8
定价：78. 00 元

电话服务
客服电话：010 - 88361066
010 - 88379833
010 - 68326294

网络服务
机 工 官 网：www. cmpbook. com
机 工 官 博：weibo. com/cmp1952
金 书 网：www. golden-book. com
机工教育服务网：www. cmpedu. com

封底无防伪标均为盗版

特别鸣谢
北京大学国家发展研究院 BiMBA 商学院
对本书提供的帮助。

Special thanks to the
BiMBA Business School of National School of
Development, Peking University for her kind help.

北京大学国家发展研究院 BiMBA 商学院
智库中的商学院

北京大学国家发展研究院

北京大学国家发展研究院（简称“北大国发院”）是北京大学一个以经济学为基础的多学科综合性学院，前身是林毅夫等六位海归经济学博士于 1994 年创立的北京大学中国经济研究中心（CCER），2008 年更名为国家发展研究院。学院秉承北大兼容并包、和而不同的学风，高度关注中国的现实问题，致力于学术与现实的结合，不遗余力地推动中国进步，是北京大学构建世界一流大学的重要组成部分。

经过近三十年的发展，北大国发院在教学、科研和智库方面都取得了非凡成就，其中又以智库的成就最为公众所熟知。这里不仅聚集了林毅夫、周其仁、张维迎、易纲、海闻、宋国青、曾毅、闵维方、汪丁丁、姚洋、黄益平、雷晓燕、徐晋涛、刘国恩、卢锋、赵耀辉、林双林、李玲、傅军、胡大源、杨壮、马浩、张黎、宫玉振等一批智囊人物和著名学者，而且对智库工作有一个共识：好的公共讨论不仅有利于形成更科学的决策，而且有利于凝聚共识，提升政策执行的效率和质量，从而更有力地推动社会进步。

北京大学国家发展研究院 BiMBA 商学院

北京大学国家发展研究院 BiMBA 商学院（原北大国际 BiMBA），是国内商学教育的翘

楚，成立于 1998 年，由北大国发院运营，汇聚全球顶级商科教育资源，国际特质鲜明，同时深得北大人文底蕴，“中西合璧、知行合一”“依国家智库、铸商界领袖”，成为中外合作办学的开创者和领导者。目前，北大国发院开设两个国际 MBA 项目：北京大学 – 伦敦大学学院（UCL）MBA 以及北京大学 – 弗拉瑞克商学院 MBA，每年秋季开学。

项目特色

全球盛誉	• 蝉联中国大陆学术水准最高商学院（QS 世界大学商学院排名）； • 中国投资回报最好的商学院，全日制 MBA、在职 MBA 连续多年名列第一（《福布斯》中文版）； • 最具市场价值的商学院（《财富》中文版）； • “一所有灵魂的商学院”（《彭博商业周刊》）。
北京大学 + 英国名校 UCL/ 欧洲名校 Vlerick 商学院，共享全球优质教育资源	• 英国伦敦大学学院（UCL），英国金砖五校之一，QS 世界大学排名位列英国第四，全球第九； • 比利时 Vlerick 商学院，欧洲最知名商学院之一，集全球 MBA 最权威三大认证：EQUIS、AACSB 和 AMBA。
中西合璧名师阵容	• 北大国发院大师云集，教学科研遥遥领先：林毅夫、周其仁、张维迎、姚洋、黄益平、雷晓燕、刘国恩、傅军、杨壮、马浩、张黎、宫玉振等； • 英、美、欧名校国际师资，教学深度结合实际，无国界探索学科前沿话题。

“知行合一的 MBA 课程体系”+丰富的选修方向，培养面向未来的复合型人才	● 国际经典 MBA 课程 + 本土创新课（Know），企业实战 + 职业导师（Do），北大精神 + 社会情怀（Be）； ● MBA 之后，可再申请选修第二学位： 全美 No. 21 的金融硕士学位（Fordham University），北京、纽约两地上课 全美 No. 1 的房地产硕士学位（The Wisconsin School of Business），美国上课。
高度与格局：国家高端智库 + 多领域研究中心	● 中国高校智库的佼佼者，站在中国及世界经济发展的高度上思考、研究并解决具体的商业问题； ● 国发院聚合北大乃至全球的研究资源，拓展交叉学科的深度与广度：中国经济研究中心、健康老龄与发展研究中心、数字金融研究中心、法律经济学研究中心、人力资本与国家政策研究中心、中国卫生经济研究中心等十几个跨学科研究中心。
完备的职业发展体系	● 全球商学院独有的“一对一职业导师”计划，指导学生成长； ● “管理创新实验室”，引领学生与企业紧密互动，围绕实际商业问题进行全景式分析和创新解决方案探讨； ● 多维度职业发展工作坊与技能讲座，全面提升职场竞争力； ● 全日制 MBA 全方位职业发展支持：自我认知、职业定位、行业洞察、企业参访、实习与就业等。
优秀多元的同窗学友	● 国发院政、商、学三界精英校友遍布全球：MBA、EMBA、DPS、EDP、经济学本硕博、CHO100、南南合作与发展学院、木兰学院等。
更多元、更丰厚的奖学金，汇聚全球优秀 MBA 学子	● 特设全额奖学金若干，特设科技英才奖学金、公益之星奖学金、一带一路奖学金、承泽英才奖学金、多元贡献奖学金、国际化人才奖学金、创业先锋奖学金、明日之星奖学金、STEM 专业人才奖学金等； ● 最多惠及 70% 的全日制学生与 30% 的在职学生。
承泽园院区	● 2021 年 9 月北京大学国家发展研究院承泽园院区建成并投入使用，以清代皇家园林为基础，新建宏伟的现代教学设施。院区设计既体现中国古典建筑之风范，又不失灵活和丰富的现代功能，园林建筑独具北大特色，并充分体现国家发展研究院深厚的文化积淀和锐意创新的精神。

项目介绍

北京大学－伦敦大学学院（UCL）MBA

北京大学和伦敦大学学院（UCL）于2016年签署了合作办学联合声明，共同开设工商管理硕士（MBA）课程项目，由北京大学国家发展研究院（NSD）与伦敦大学学院管理学院（SoM）共同运营。北京大学国家发展研究院是领先的经济学和公共政策智库，而伦敦大学学院管理学院是领先的专注于创新（innovation）、创业（entrepreneurship）、技术（technology）和分析（analytics）的商学院。因此，该项目是北京大学和伦敦大学学院结合双方专长而开设独特课程的一个绝佳典范，将对培养下一代领袖人才、推动中国发展和向知识经济转型产生重要作用。英国伦敦大学学院（University College London）简称UCL，建于1826年，与剑桥大学、牛津大学、帝国理工学院、伦敦政治经济学院并称“G5超级精英大学”，代表了英国最顶尖的科研实力、师生质量、经济实力，同时也是英国顶尖研究型大学联盟——罗素大学集团（包括24所成员大学，类似于美国的“常春藤”联盟）的成员，享有英国政府最多的财政预算。UCL的研究能力位列英国第二（来源：Research Excellence Framework (REF) 2021），其在2024QS世界大学排名中，位列全球No.9；在2024《泰晤士报》世界大学排名中，位列全球No.22。

扫码了解更多项目信息

北京大学－弗拉瑞克商学院MBA

北京大学国家发展研究院（NSD）与比利时弗拉瑞克商学院（Vlerick Business School）自2008年起合作创办北大－Vlerick MBA。比利时是欧盟核心的行政办公所在地，也是整个欧洲地理和文化的交汇之地，被誉为欧洲之都，是观察和体味欧洲经济、政治、

扫码了解更多项目信息

文化的重要窗口，也是国际交流的重要枢纽。弗拉瑞克商学院是欧洲历史最悠久商学院，由鲁汶大学和根特大学于 1953 年共同建立，是全球为数不多能同时拥有管理教育领域三大国际认证（EQUIS、AACSB、AMBA）的商学院。弗拉瑞克商学院在英国《金融时报》欧洲 MBA 排名中位列 No.23，在 QS 全球 MBA 排名中位列 No.62。

北大 -Vlerick MBA 项目不仅是北大与 Vlerick 师资的珠联璧合，还汇集了来自美国加州大学伯克利分校 Hass 商学院、约翰霍普金斯大学 Carey 商学院、法国 INSEAD 商学院、ESSEC 商学院等全球名校的师资。一流的师资与课程将带给学生经典理论与创新方法，融汇全球化与本土化，践行知成一体，培养有高度、有视野、有格局的未来企业家和高级管理人才。该 MBA 项目以全英文授课，设置在职 MBA 班。为使 MBA 同学更好地面向实践，面向创新，面向未来，北大 -Vlerick MBA 项目为学生设置全球商学院独有的三导师制，即“学术导师 + 企业导师 + 职业导师”，和“管理创新实验室”的商科实践性项目。三导师制的设定，一方面通过学术导师与企业导师的帮助，使得 MBA 学生把商业理论和企业实践有机结合；另一方面，在职业导师的带领下，同学们可以更好地探寻出一条充分发挥自我潜质和比较优势的职业方向与路径。“管理创新实验室”项目汇集来自各行各业的企业和企业家导师，尤其是科创领域的企业与企业家。通过此项目的历练，MBA 学生可以加强其洞察科技新方向和把握创新基本点的能力，从而活学善用，知成一体，成长为有视野、谋布局、善创新、懂市场的高级管理者和引领时代的商业领军者。

联系我们

电话：010-62754800

邮箱：admissions@bimba.pku.edu.cn

网址：www.bimba.pku.edu.cn

地址：北京市海淀区蔚秀园路北京大学国家发展研究院承泽园办公楼

微信群：如希望加入申请咨询群，请添加管理员微信 PKUNSD-BiMBA。

ChaseDream 创始人寄语

批判性推理算得上是 GMAT 考试语文部分最有意思的一项了，它经常被同学们称为“逻辑”。但是，GMAT 考试中的这个“逻辑”和生活中的逻辑大不相同。在日常生活中，我们所说的“你这个人说话真有逻辑”里的“逻辑”多半是“有理有据”的意思。而在 GMAT 考试中，“逻辑”的意思多半是让我们对一个段落进行批判性思考，继而评估这个段落。

我见过很多同学只是凭借选项和段落“是否有关系”这种感觉来做逻辑考题，这种“靠不住”的感觉让他们的正确率总是上下波动，很不稳定。为什么说这种感觉“靠不住”呢？因为，你对 GMAT 逻辑题的“感觉”一小部分是来源于你个人的天赋超群，骨骼清奇，而绝大部分是来源于“题海”。通过夜以继日的做题和总结，终于找到一种所谓的“GMAT”感觉。但是，GMAT 不会考查重复的题目，如果你没有掌握公式定理，仅仅是对现有题目进行现象总结，那么做多少题也难以保证你在考场上所向披靡。

那么，为什么本书中所阐述的解题方法是“靠得住”的呢？毕出的这种解题方法是来源于逻辑学的，他参考并研究了许多有关普通逻辑学、非形式逻辑学、形式逻辑学等方面的专业书籍，将这些学科与 GMAT 考试进行匹配，并且用简练的语言和文字以解题公式的形式呈现给大家，真正做到了从根本上理解 GMAT 逻辑的考查要点，正面攻破 GMAT 逻辑，杜绝了“从答案倒推解法”形成“ GMAT 感觉”的现象。在研究的同时，毕出还经常和命题官方讨论求证，把握命题思路并且日益完善解题方法。因此，一旦学会这本书中的“公式定理”，你将在 GMAT 考试中以不变应万变。

GMAT 是一个设计精巧的考试，通过各种方法都有可能拿到很高的分数。或许这本书不是通向 GMAT 高分的唯一途径，但是，如果你希望少走弯路或当你试过多种方法却依然没有头绪的时候，不妨试试本书中的方法。相信它至少能帮你从容地躲过一些荆棘，迈过一些关卡，更为轻松地到达成功的彼岸。

Zeros

ChaseDream Founder

2024 年 2 月

ChaseDream 总编推荐序

批判性推理（Critical Reasoning，简称为 CR），似乎从未被我们作为一门必要的课程系统地学习过。

回想我久远的学生时代，能和逻辑沾边儿的，好像只有历任数学老师都会强调的那句话："数学，可以帮助你训练缜密的逻辑思维。"说完这句话，老师经常会推一推鼻梁上的眼镜，脸上划过一丝骄傲的微笑。至于如何训练、训练了哪些方面，我无从知晓，好像当时也没有关心过。

这使得很多人在第一次接触 GMAT CR 类型的题目时，完全不知所措。甚至那些自认为逻辑很好的同学，看着题目和正确答案，脑补了一堆奇奇怪怪的东西，也无法还原题目的推理过程。

不知道是否与此有关，市面上绝大部分的 GMAT CR 复习指导书，都并没有真正系统地讲解批判性推理。更多的时候似乎只是单纯地按照题目的问法来进行分类：削弱、支持、假设、评估等，再根据官方的样题，总结出一些解题技法，告诉你如何削弱、如何支持、怎样假设、怎样评估。这些做法看似正确，却没有道出 CR 题型的考查点。而题型分类方法造成的偏差，也让考生们很难建立正确、完整的批判性推理体系，从而在考场上无所适从。

所以，在一切的开始，还是让我们利用 GMAT CR 考题，小小地补一堂批判性推理的课吧。经过学习，你不但可以在 GMAT 考试中从容获取高分，还会在未来商学院的学习中更加轻松。甚至，你还会惊喜地发现，学完本书以后，就算和别人“吵架”时你都可以站在另一个高度藐视对方，轻松指出对方观点中的各种逻辑错误。

好了，翻过这页，你将看到解决 GMAT CR 题目的真正钥匙。

少年，Go ahead！

steven

ChaseDream 总编

前　言

当今世界，批判性思维是决策的关键能力，无论是在商业、法律还是日常生活中。它启迪我们辨识逻辑漏洞、加强论点，并提出坚实的结论。本书的目的正是帮助大家建立并精炼这个能力，尤其适合备战 GMAT 考试中批判性推理部分。

在第一章中，我们提供了批判性推理的基础，从演绎论证和假说论证的结构开始，探索逻辑运算符、模态命题、量词等构成有效论证的基本元素。大家将了解到如何系统地构建论证，并学会如何辨别一个论证是否站得住脚。

第二章深入探讨了 GMAT 批判性推理部分的四大题型，为大家提供了一系列的解题策略和技巧，旨在帮助大家提高分析论证的精准度、构建有说服力的论证、批判性评价他人的推理，以及有效地规划解题方法。这不仅仅是为了考试，更是为了训练思维的灵活性和深度。

第三章包含了多组综合提高训练，旨在帮助大家通过实践加深理解，并将书中提及的概念运用到 GMAT 批判性推理题型类似的复杂情境中。这些训练均经过精心设计，以模拟实际考试环境，帮助大家在实际考试中更加从容不迫。

本书将成为您批判性推理训练的指南和伙伴。通过本书的学习，您将获得一个更加敏锐和结构化的思考方式，为 GMAT 考试和未来的挑战做好准备。欢迎您开始这段深化逻辑与推理技巧的旅程。

由于时间仓促，水平有限，本书难免会有不足和纰漏之处，欢迎热心而诚恳的读者批评指正，并且对本书提出宝贵意见。意见请发送至：book@chasedream.com，感谢斧正。

毕　出

于2024年春

目 录

02 第二章 GMAT 批判性推理考题的四大题型 // 023

03 第三章 综合提高训练 // 241

第一章

GMAT 批判性推理概述

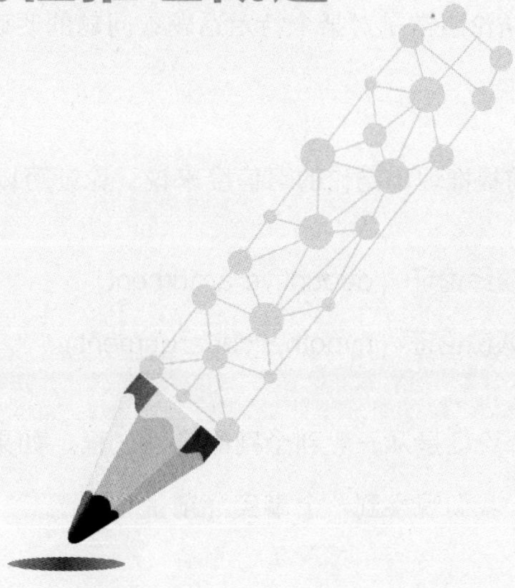

批判性推理是一门研究论证逻辑的学科。论证是一种逻辑推理过程，其目的是通过提供理由或证据来支持或反驳某个主张、观点或结论。在论证中，通常包括两个主要部分：前提（premise）和结论（conclusion）。

前提：

前提是论证的基础，提供支持结论的理由或证据。

前提可以是事实、观点、数据、统计信息或其他形式的陈述，其目的是为结论提供支持。

结论：

结论是论证的核心陈述，是需要被证明或支持的观点。

结论通常是对某个特定话题或问题的明确观点、判断或建议。

从前提推理出结论的可信度来说，论证可以分为两大类：

1. 演绎论证（deductive argument）
2. 假说论证（hypothetical argument）

演绎论证是从一般到个别的推理过程。如果前提正确，则结论在逻辑上必然正确。例如，我们观察到一个事实（前提）：

（1）小明扶着老奶奶过马路。

如果只根据演绎论证的逻辑进行推理，则我们可以得到下列结论：

（2）存在小明这个人。

（3）小明也通过了那条马路。

显然，如果句（1）是正确的，那么句（2）和句（3）必然是正确的。

假说论证是从现象到假说的推理过程。如果前提正确，则结论在逻辑上可能是有道理的，但并非一定正确，依然以句（1）为前提，我们可以得到下列结论：

（4）小明是个乐于助人的人。

（5）老奶奶会对小明表示感谢。

显然，即便句（1）正确，句（4）和句（5）也不一定正确，它们仅是可能的假说。

1.1 演绎论证（Deductive Argument）

1.1.1 演绎论证的基本规则

演绎论证的结构通常比较固定。它从一个或多个普遍性的前提出发，得出特定情况下的必然结论。下面请看一个更正式的例子：

前提 1：小明每周三一定会穿衬衫。

前提 2：今天是周三。

结　论：小明今天会穿衬衫。

在该例子中，如果两个前提同时正确，则结论必然正确。在 GMAT 考试中，这种类型的考题考法是比较多样的。最简单的考法是：它可以在题干中给出前提，让我们选择合适的选项作为结论。例如：

小明每周三一定会穿衬衫。今天是周三。

如果上述说法正确，那么以下哪个说法一定正确？

（A）小明今天会穿衬衫。

（B）小明喜欢穿衬衫。

例题答案为选项 A 。显然，选项 B 是不一定成立的，即我们无法确定小明是否喜欢穿衬衫。

此外，GMAT 考题也可以“删除”某个前提，让我们选择合适的选项作为该论证的“假设”，例如：

小明每周三一定会穿衬衫。因此，小明今天会穿衬衫。

上述论证基于下列哪项假设?

(A) 今天是周三。

(B) 小明喜欢穿衬衫。

例题答案为选项 A 。显然，选项 A 补全了推理文段，使之成为“演绎论证”。

演绎论证还可以与方案相结合，例如：

通常来说，增加广告支出可以提高产品知名度，从而吸引更多消费者。改进产品质量会提高客户满意度和忠诚度，进而带来更多的重复购买和口碑推荐。

以下哪个方案可以帮助公司实现市场份额的增加?

(A) 公司应该增加广告支出，并专注于提高产品质量。

(B) 公司应该积极招聘，尽力网罗天下英才。

例题答案为选项 A 。显然，选项 A 提出的方案是可以通过推理文段演绎推理得出的，即如果推理文段的信息正确，则该方案必然可以帮助公司增加市场份额。

演绎论证要求我们充分理解前提和结论，掌握其中明示与暗示的信息，做出严谨的推理和判断。

1.1.2 模态命题

所谓模态命题，其实就是陈述事物情况的必然性或可能性的句子。例如，“小明必然去买了三本书”和“小明可能去买了三本书”。

类似于这两者的关系，我们直接按照常识来理解即可。

显然，从“小明必然去买了三本书”推理出“小明可能去买了三本书”这一结论是符合演绎论证规律的，反之则不符合演绎论证规律，即如果“小明可能去买了三本书”是一个事实，那么我们无法得知小明是否必然会去买书。

英语中有一些表达可能性的方式，如下表：

Necessity	Probability	Possibility
certainly	probably	can
clearly	likely	could
definitely	more likely than not	may
must		maybe
necessarily		might
surely		perhaps
		possibly

从左至右的推理都被视为是有效的，即由 Necessity 可以推理出 Probability，也可以推理出 Possibility；从右至左的推理都被视为是无效的，即不能由 Possibility 推理出 Probability，更不能推理出 Necessity。

1.1.3 逻辑运算符

逻辑运算符（logical operator）是用于构建逻辑表达式的符号，它们用于表示逻辑关系和进行逻辑运算。最常用的逻辑运算符包括：

与（AND）：

它表示两个陈述都必须为真，整个表达式才为真。以下单词和词组经常被用来表达陈述 A 和陈述 B 之间的“与”关系：

A and B	A，even though B	not only A but also B
Although A，B	A，whereas B	A but B
A. However，B	A. Furthermore，B	

例如，

"虽然小明是好人，但他也是打了小李的人。"这句话如果是真的，那么它意味着：

（1）小明是好人。

（2）小明打了小李。

（1）和（2）均为真。

或（OR）：

它表示两个命题中至少有一个为真时，整个表达式就为真。它通常用 A or B，either A or B 和 A unless B 来表示。

例如，

"小明是好人，或者他打了小李。"这句话如果是真的，那么它意味着：

（1）小明是好人。

（2）小明打了小李。

（1）或（2）有一个陈述为真。

非（NOT）：

它是一个一元运算符，用于反转一个命题的真值。

例如，

> 如果 A 为真，则非 A 为假；如果 A 为假，则非 A 为真。

蕴含（IMPLIES）：

它表示如果第一个命题为真，则第二个命题也必须为真。GMAT 考试中此类运算符通常用“条件性陈述”的方式来体现。以下单词和词组经常被用来表达陈述 A 和陈述 B 之间的“条件”关系：

A would mean that B	B if A	A，only if B
If A，then B	Not A unless B	B provided that A

例如，如果下雨（A），那么地面会湿（B）。

这种陈述建立了两个命题之间的条件关系：一个是条件（如果部分），另一个是结果（那么部分）。

在日常语言中，条件性陈述通常暗示着因果关系或者逻辑上的依赖。

其真值表的特殊之处在于：

> 当 A 为假时，无论 B 是否为真，这个条件命题依旧为真。

让我们看一个更贴近生活的例子：

假设我们打赌，赌注为一顿饭。

我说："如果你 GMAT 考到了655 分以上，那么我就会请你吃饭。"

请问，在什么情况下，我违背了赌约呢？

只有一种情况，即你 GMAT 考到了655 分以上，但我没有请你吃饭。

因此，如果"你没考到655 分以上"（陈述 A 为假），无论我是否请你吃饭了（无论陈述 B 是否为真），我都不算违背赌约（这个条件命题依旧为真）。

"蕴含"这一运算符在许多 GMAT 考题中均有体现。例如，下面这句话：

如果小明努力，那么小明就能成功。

我们是否可以说"出于某种原因，小明可能很难做到'努力'这件事"呢？当然是不能的。

要记住，仅当条件为真且结果为假时，这个句子才是"假"的。因此，考虑条件是否为假，完全无法质疑整个命题。

例题 1

Early in the twentieth century, LakeKonfa became very polluted. Recently fish populations have recovered as release of industrial pollutants has declined and the lake's waters have become cleaner. Fears are now being voiced that the planned construction of an oil pipeline across the lake's bottom might revive pollution and cause the fish population to decline again. However, a technology for preventing leaks is being installed. Therefore, provided this technology is effective, those fears are groundless.

The argument depends on assuming which of the following?

(A) There is no reason to believe that the leak-preventing technology would be ineffective when installed in the pipeline in LakeKonfa.

(B) The bottom of the lake does not contain toxic remnants of earlier pollution that will be stirred into the water by pipeline construction.

该例题的结论为：

Provided this technology is effective, those fears are groundless.

这是一个非常典型的蕴含关系。如果这句话是真的，那么它必然假定了“在技术有效的情况下，安装输油管不会产生污染”（命题）必然为真。所以，选项 B 是正确的。对于选项 A 来说，蕴含关系从不要求“条件”一定成立。反过来说，我们对条件部分的质疑或加强均无法对整个蕴含关系产生影响。

1.1.4 演绎论证中的量词

例如，

(1) 有些 A 是 B。

(2) 所有 A 都是 B。

句 (1) 中的“有些”和句 (2) 中的“所有”都是量词。在演绎论证的考题中，当推理文段或者选项中出现大量诸如 all, some, any, no 等这些“量词”时，我们就可以判断此题是在考查“对当关系”。英语中常见的量词见下表：

全部	大部分	有些	没有
all	generally	a number	never
always	a majority	a portion	no
any	most	at least one	none
both	more than half	occasionally	not any
each	usually	one or more	not one
every		some	nowhere
everywhere		sometimes	
whenever		somewhere	
wherever			

对于这类考题，解题方法是将推理文段中有意义的信息全部转换成命题形式：

（3）所有A都是B。

（4）有的A是B。

（5）所有A都不是B。

（6）有的A不是B。

根据上面这四种形式，我们绘制出韦恩图即可求解。句（3）（4）（5）（6）对应的韦恩图分别为（橙色圈为A，黑色圈为B，下图主要体现两个集合间的关系而非大小比较）：

句（3）：

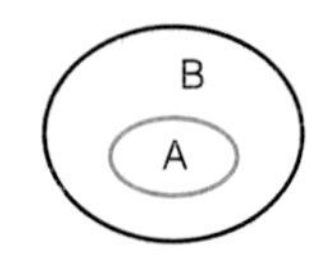

句（4）：

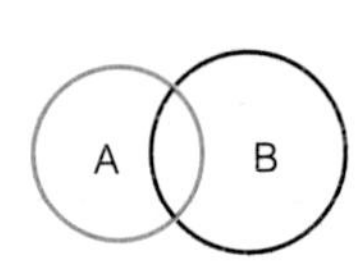

句（5）：

句（6）：

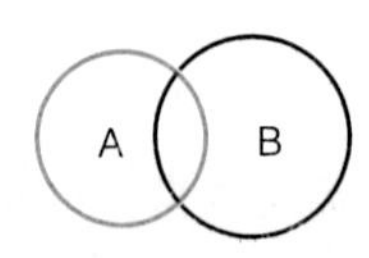

请注意，条件性陈述也可以转化为简单命题。例如：

如果小明打了小李，那么他就是坏人。

这句话蕴含的逻辑关系是：

所有打了小李的人都是坏人。

1.2 假说论证（Hypothetical Argument）

1.2.1 假说论证的基本规则

假说论证的结构比较多样，是从现象到假说的推理过程。如果前提正确，则结论在逻辑上可能是有道理的，但并非一定正确。我们可以沿用第一章开篇给出的例子：

前提：小明扶着老奶奶过马路。

结论：小明是乐于助人的人。

在该例中，如果前提正确，则结论不一定正确，毕竟小明可能仅仅是作秀，并不是真心帮助那位老奶奶。

在 GMAT 考试中，这种类型的考题考法也十分多样。最常见的方法是让我们加强、削弱或评价一个假说论证，例如：

小明扶着老奶奶过马路。因此，小明是乐于助人的人。

以下哪个选项，如果正确，能最强地削弱上述推理？

（A）小明喜欢作秀，这次扶老奶奶过马路的行为是精心设计过的。

（B）是否乐于助人是衡量一个孩子好坏的重要标准。

例题答案为选项（A）。这类考题的本质是在考查我们对于假说论证中可能出现的漏洞（在逻辑上称之为谬误）的把握。

要想准确评估一个假说论证的有效性，我们首先要了解假说论证的基本类型：

（1）普通预测推理

（2）泛化推理

（3）类比推理

（4）归因推理

（5）因果推理

1.2.2 普通预测推理

普通预测推理是最常见的假说论证类型，它通常基于一定经验以及常识性的因果关系。这类推理的前提通常是一个现象，结论则是作者基于这个现象对未来做出的预判或对其可能产生的结果的叙述。例如：

前提：A 公司成本增加。

结论：该公司的利润将会下降。

又例如：

前提：房子因地震而坍塌。

结论：小明将会受伤。

显然，这两个例子的结论都是“有道理”的，但它们并非一定正确。究其本质，这两个例子的前提均不是能令结论成立的充分条件。

（注：通俗来说，“充分条件”的意思是：只要达成了该条件，结论就必然会发生。）

在第一个例子中，利润下降的充分条件是“A 公司的总收入和总成本的差值下降”；在第二个例子中，小明受伤的充分条件是“房子坍塌砸到了小明（且达到了让小明受伤的程度）”。

1.2.3 泛化推理

泛化推理通常使用样本（前提）来支持总体（结论）。例如：

> 前提：小明学习成绩很好。
>
> 结论：小明所在班级中所有同学的学习成绩都很好。

显然，“小明”是样本，“小明所在班级中所有同学”是总体。这种推理是非常直白的，结论仅仅把前提的范围扩大。

评价一个泛化推理的好坏主要有两个维度。

（1）样本是否存在偏差。

顾名思义，我们要考虑用来支持总体的样本是否具有足够的代表性。例如，上面例子中应该考虑“小明”是否是班级里智力水平较为出众的一个。

> **示例 1**
>
> 市场研究员：我在一个著名的在线论坛上看到大多数人都不喜欢最新款的智能手机。因此，这款手机肯定是市场上大部分人都不喜欢的。
>
> 谬误分析：这个论证假设在线论坛的用户代表了所有智能手机的潜在用户。然而，论坛用户可能只代表了一个特定的群体，他们的观点可能与广大消费者的看法不同。例如，论坛用户可能更关注技术细节，而一般消费者可能更关注价格或品牌。因此，从论坛用户的观点推断出整个市场的态度是有偏差的。

示例 2

我们在大学校园内进行了一项调查，发现 90% 的年轻人支持某政治候选人。因此，我们可以推断这位候选人在全国范围内压倒性拥有年轻选民的支持。

谬误分析： 这个论证假设大学校园内的调查结果可以推广到所有年轻人。然而，大学生可能在多个方面（如教育水平、政治观点、社会经济背景）与其他年轻人群体有显著差异。因此，仅凭校园调查结果推断全国年轻人的政治倾向是具有偏差的。

（2）样本数量是否足够。

即便前提是一个足够好的样本，它也依然可能是单薄的，我们还需要考虑是否应找到更多的样本。例如，本小节刚开始的例子中应该考虑小王是否成绩也不错。

示例 3

我上周去了一个北方城市度假，那里的天气非常冷。这说明那个城市一年四季都非常寒冷。

谬误分析： 这个论证基于短期的个人经验（一周的度假）来判断一个地区的整体气候。这忽视了天气的季节性变化和长期气候模式，因此，样本数量是不足的。

示例 4

我们公司最近采用了一种新技术，但我的一个同事因为不熟悉这项技术而导致工作效率下降。因此，这项新技术对提高工作效率没有帮助。

谬误分析： 这个论证仅基于一个员工的经验就对新技术的整体效益做出了结论。它没有考虑其他员工的适应情况或长期效益，因此，样本数量是不足的。

1.2.4 类比推理

类比推理，首先提到两个或两个以上的事物在某些方面是相似的，然后将关于一个事物的主张作为接受关于另一个事物的相似主张的理由。例如：

前提：小李去吃饭。

结论：小明去吃饭。

因为类比推理的前提和结论在“类比对象”的相似度不够时是完全无法契合的，甚至是完全没有道理的，所以在评估一个类比推理时，需要考虑是否存在“其他相关相似点缺失（the absence of additional relevant similarities）”。例如，我们需要考虑“小李”和“小明”是否是在相同的时间吃了上一顿饭等。

示例 1

前提：电脑能够存储大量信息并快速处理复杂运算。

结论：人脑也能够存储大量信息并快速处理复杂运算。

谬误分析：人脑和电脑在处理信息方面的原理可能有所不同。

示例 2

前提：在太阳系中，行星围绕太阳转动。在原子模型中，电子围绕原子核转动。

结论：太阳系可以作为理解原子结构的一个类比，其中太阳类似于原子核，行星类似于电子。

谬误分析：太阳系和原子结构在受力模式上可能有所不同。

1.2.5 归因推理

归因推理也是一种常见的假说论证方式，其核心在于探究前提中现象的成因，或寻找解释该现象发生的原理（动机）。例如：

前提：家里被翻找得十分杂乱。

结论：可能是有小偷来过了。

在这个例子中，结论在探究为何“家里会很杂乱”，所以是典型的归因推理。又例如：

前提：小明扶着老奶奶过马路。

结论：小明是乐于助人的人。

在这个例子中，结论在试图寻找“小明会扶着老奶奶过马路”这一行为的原因（背后动机）。

评估归因推理有两个维度：

（1）是否有其他证据

结论的成立还需依赖是否存在其他证据。例如，上面两个例子需要分别考虑“家里是否有门窗被撬动的痕迹”和“小明是否有其他乐于助人的行为举动”。

（2）是否有其他可能的原因和解释

除结论外，是否还有其他成因。例如，上面两个例子需要分别考虑“是否是家里有亲戚来过但忘记整理了”和“这次扶老奶奶过马路的行为是否是精心设计过的”。

1.2.6 因果推理

有些假说论证是基于对事件之间因果关系的洞察而做出的。因果推理在本质上具有挑战性，因为它很难被直接观察到。另外，针对因果的本质，也没有科学或者哲学的共识，更多凭借经验。例如：

前提：我们观察到，凡是考 GMAT 的人都聪明。

结论：考 GMAT 导致人聪明。

上例中，前提展示“考 GMAT”和“聪明”这两个事件经常同时出现（或先后出现），结论中直接给出了这两件事是具备因果关系的。

可以这样说，凡是前提展示两个事件同时或先后发生并且结论展示了这两个事件的因果关系的推理都是因果推理。请再看以下两个示例，从中体会因果推理的特点。

示例 1

前提：在一项长期研究中，经常运动和均衡饮食的人群患心脏病的比例较低。

结论：经常运动和均衡饮食可能导致心脏病发病率降低。

示例 2

前提：统计数据显示，拥有高等教育学历的人平均收入高于没有高等教育学历的人。

结论：高等教育可能导致个人拥有更高的收入水平。

评估因果推理有三个维度：

(1) 纯粹巧合

事件 A 和事件 B 之间不存在因果的原理，只是纯粹在时间或空间上巧合罢了。例如：

前提：凡是考 GMAT 的人都聪明。

结论：考 GMAT 导致人聪明。

上例中，“考 GMAT” 和 “聪明” 在前提中同时出现，在结论中呈现因果关系。若想削弱该推理，我们可以说：

“GMAT 只是一个考试，它不会改变人脑的生理结构。”

这个说法指出 “考 GMAT” 和 “聪明” 是纯粹巧合的，即事件 A 和事件 B 之间不存在因果的原理。当然，如果想让结论的可信度更高，我们可以给出因果的原理或者增加其他能确认因果关系的证据。

请再看以下两个示例，从中体会纯粹巧合的特点。

示例 1

前提：在一项长期研究中，经常运动和均衡饮食的人群患心脏病的比例较低。

结论：经常运动和均衡饮食可能导致心脏病发病率降低。

谬误分析：“经常运动和均衡饮食” 与 “是否患心脏病” 可能根本不存在因果关系。在医学上，“经常运动和均衡饮食” 与 “是否患心脏病” 可能是两个完全独立的变量。另外，这个示例仅基于某一项长期的研究，我们还可以给出其他发现 “经常运动和均衡饮食” 与 “是否患心脏病” 不存在相关性的研究来削弱推理。

示例 2

前提：统计数据显示，拥有高等教育学历的人平均收入高于没有高等教育学历的人。

结论：高等教育可能导致个人拥有更高的收入水平。

谬误分析："拥有高等教育学历"与"收入水平"可能根本不存在因果关系，只是在统计数据中恰好同时出现。高等教育学历或许只是作用在学术方面，而收入则可能不由学术造诣决定。

(2) 因果倒置

也许是事件 B 导致事件 A，而非事件 A 导致事件 B。例如：

前提：凡是考 GMAT 的人都聪明。

结论：考 GMAT 导致人聪明。

若想削弱该推理，我们可以说：

"是那些本身就聪明的人才会去考 GMAT。"

若想让结论更加可信，我们可以排除因果倒置的可能，即：

不是那些本身就聪明的人才去考 GMAT。

请再看以下两个示例，从中体会因果倒置的特点。

示例 1

前提：在一项长期研究中，经常运动和均衡饮食的人群患心脏病的比例较低。

结论：经常运动和均衡饮食可能导致心脏病发病率降低。

谬误分析：有一种可能性，是那些本身心脏就比较好的人才会喜欢去运动。

示例 2

前提：统计数据显示，拥有高等教育学历的人平均收入高于没有高等教育学历的人。

结论：高等教育可能导致个人拥有更高的收入水平。

谬误分析：或许是收入水平的提高导致很多人想进一步提升自己，然后才完成了更高等的教育，并且获得了学历。

（3）他因导致结果

也许存在另一个因素——事件 C 导致了事件 B 的发生，又或者，这个事件 C 同时导致了事件 A 和事件 B 的发生（请注意，此时的事件 C 应和事件 A 相矛盾）。例如：

前提：凡是考 GMAT 的人都聪明。

结论：考 GMAT 导致人聪明。

若想削弱该推理，我们可以说：

“考 GMAT 的人都经常吃脑黄金，而这个东西是可以提高智力水平的。”

若想让结论更加可信，我们可以排除他因导致结果的可能，即：

不存在考 GMAT 的人都会吃脑黄金的情况。

请再看以下几个示例，从中体会他因导致结果的特点。

示例 1

前提：在一项长期研究中，经常运动和均衡饮食的人群患心脏病的比例较低。

结论：经常运动和均衡饮食可能导致心脏病发病率降低。

谬误分析：这个推理可能过度简化了心脏病发病率的原因。心脏病的发生可能受多种因素的影响，包括遗传因素、环境因素、心理压力等。仅从运动和饮食的角度来解释心脏病发病率的变化可能忽略了这些其他重要因素。

示例 2

前提：统计数据显示，拥有高等教育学历的人平均收入高于没有高等教育学历的人。

结论：高等教育可能导致个人拥有更高的收入水平。

谬误分析：这个推理可能过度简化了收入水平高的原因。收入水平可能受多种因素的影响，包括家庭因素、时代红利、行业因素等。

第二章

GMAT 批判性推理考题的四大题型

从本章开始，我们将开始学习 GMAT 批判性推理考题的四大题型。分别为：

(1) 分析论证
(2) 构建论证
(3) 评估论证
(4) 方案

2.1 分析论证（Analysis）

分析论证要求我们理解文章的逻辑结构以及每句话扮演的角色。在论证中，两个最重要的角色是“前提”和“结论”。

前提：结论基于的证据，是能推理得出结论的句子。

结论：作者通过推理最终得出的观点句。

通过这个定义可知，结论一定是作者最终“猜”出来的，所以它不能是事实。试想，如果某个结论已经是事实了，那就不用推理了，因为存在即合理。反之，前提一定是事实，不可也无须被质疑。

2.1.1 找准结论的方法

找准结论主要依靠两种方法：

1. 结论信号词

通常，结论的前面会出现一个信号词。当出现这类信号词时，后面必然出现该推理文段的结论。这类信号词包括：therefore，consequently，clearly，thus 等。下面例题中用了结论信号词 clearly。

Because it was long thought that few people would watch lengthy televised political messages, most televised political advertisements, like commercial advertisements, took the form of short messages. Last year, however, one candidate produced a half-hour-long advertisement. During the half hour the advertisement was aired, a substantial portion of the viewing public tuned into the advertisement. Clearly, then, many more people are interested in watching lengthy televised political messages than was previously thought.

Clearly 之后为结论句部分。

2. 被支持的句子

有些考题没有结论信号词，抑或是我们人为地将例题中的 clearly 去掉了，例如：

Because it was long thought that few people would watch lengthy televised political messages, most televised political advertisements, like commercial advertisements, took the form of short messages. Last year, however, one candidate produced a half-hour-long advertisement. During the half hour the advertisement was aired, a substantial portion of the viewing public tuned into the advertisement. Many more people are interested in watching lengthy televised political messages than was previously thought.

进行如此操作后，结论句依然不变。由此可知，信号词本身并不影响我们对于结论句的选取。实际上，若题目中没有结论信号词，我们就需要从语义上判断结论。在整个推理文段中，一定有且仅有一句话是被其他句子所支持的，而其他的句子都是为这句话服务的。这个被支持的句子就是“结论句”。

从经验上讲，很多题目的结论句都出现在推理文段的末尾。因此，请对末尾的句子更加留心。

2.1.2 剖析出“真正”的结论

GMAT 试题为了让推理文段更贴近生活，它有时会用比喻等修辞手法来呈现结论。此时，为了能更准确地理解结论，我们需要根据上下文，剖析出真正的结论，例如：

为了能使牛肉的味道更好，很多养牛的商人会给牛棚装上音乐播放器。但是，牛对于音律没有任何的感知能力，因此，给牛放音乐是在浪费生命。

结论是：给牛放音乐是在浪费生命。

根据上下文可知，结论的真正意思是：给牛放音乐对牛没有任何帮助（至少对牛的肉质没有任何提升）。

我们来看一道英文原题。

Often patients with ankle fractures that are stable, and thus do not require surgery, are given follow-up X-rays because their orthopedists are concerned about possibly having misjudged the stability of the fracture. When a number of follow-up X-rays were reviewed, however, all the fractures that had initially been judged stable were found to have healed correctly. Therefore, it is a waste of money to order follow-up X-rays of ankle fracture initially judged stable.

推理文段的结论很明确：...it is a waste of money to order follow-up X-rays of ankle fracture initially judged stable.

根据上下文可知，该结论真正的意思是：在任何情况下，所有做第一次 X 光检查后被诊断为稳定的骨折都能完全康复。

这是因为，但凡有一例病人做第一次X光检查后的诊断不准，那么做第二次X光（follow-up）检查就不能说是浪费钱。毕竟，千分之一甚至万分之一的失误概率在医学上可能也意味着一条人命。

2.1.3 分析论证类考题的解法

分析论证类的考题问法包括但不仅限于：

（1）In the argument given, the two portions in boldface play which of the following roles?

（2）The passage above proceeds by...

（3）Jamal responds to Sasha by doing which of the following?

（4）The statements above most strongly suggest that the main point of disagreement between the critics and the spokesperson is whether...

这类考题要求我们充分理解文章，知晓每一个句子在论证中的作用（是前提还是结论），同时了解评价“假说论证”的各种维度，并且可以识别出说话者是在哪个维度上评估论证的。

当遇到有黑体句的考题时，解题要抓住两个要素：

（1）通读整个文段，判断文段的主结论；确定主结论后，判断黑体句和主结论的关系。通常来说，黑体句和主结论的关系有四种，分别为：和主结论一致、和主结论相矛盾、和主结论无关、本身就是主结论。

（2）看选项时，不要过分关注那些抽象名词的意思（例如下面例题1选项B中的finding和选项D中的evidence等），要把关注点放在这些抽象名词后面的定语从句上。那些定语从句真正揭示了黑体句的作用。

例题 1

Hunter：Many people blame hunters alone for the decline in Greenrock National Forest's deer population over the past ten years. **Yet clearly, black bears have also played an important role in this decline.** In the past ten years, the forest's protected black bear population has risen sharply, and examination of black bears found dead in the forest during the deer hunting season showed that a number of them had recently fed on deer.

In the hunter's argument, the portion in boldface plays which of the following roles?

(A) It is the main conclusion of the argument.
(B) It is a finding that the argument seeks to explain.
(C) It is a correct explanation that the argument concludes.
(D) It provides evidence in support of the main conclusion of the argument.
(E) It introduces a judgment that the argument opposes.

确定文段的主结论并判断黑体句与主结论的关系。

本文段的主结论为：...black bears have also played an important role in this decline.

黑体句本身就是该文段的主结论。

选项分析

(A) 正确。它是推理文段的主结论。
(B) 它是一个推理文段寻求解释的发现。推理文段没有尝试去解释这个黑体句，而是在反驳该黑体句，排除。
(C) 它是一个推理文段总结下来的正确的解释。推理文段不同意这个黑体句所叙述的观点，排除。

(D) 它给出了一个支持推理文段主结论的证据。黑体句不是一个证据，而是一个有待验证的判断，排除。

(E) 它提出了一个推理文段反对的判断。黑体句本身就是主结论。

例题 2

As a large corporation in a small country, Hachnut wants its managers to have international experience, **so each year it sponsors management education abroad for its management trainees**. Hachnut has found, however, that the attrition rate of graduates from this program is very high, with many of them leaving Hachnut to join competing firms soon after completing the program. Hachnut does use performance during the program as a criterion in deciding among candidates for management positions, but **both this function and the goal of providing international experience could be achieved in other ways**. Therefore, if the attrition problem cannot be successfully addressed, Hachnut should discontinue the sponsorship program.

In the argument given, the two boldfaced portions play which of the following roles?

(A) The first describes a practice that the argument seeks to justify; the second states a judgment that is used in support of a justification for that practice.

(B) The first describes a practice that the argument seeks to explain; the second presents part of the argument's explanation of that practice.

(C) The first introduces a practice that the argument seeks to evaluate; the second provides grounds for holding that the practice cannot achieve its objective.

(D) The first introduces a policy that the argument seeks to evaluate; the second provides grounds for holding that the policy is not needed.

(E) The first introduces a consideration supporting a policy that the argument seeks to evaluate; the second provides evidence for concluding that the policy should be abandoned.

确定文段的主结论并判断黑体句与主结论的关系。

本文段的主结论为：If the attrition problem cannot be successfully addressed, Hachnut should discontinue the sponsorship program.

第一个黑体句算是背景，它给出的方案是主结论不同意的，即和主结论相矛盾；第二个黑体句在支持主结论，即和主结论方向一致。

选项分析

(A) 第一个黑体句描述了论证试图去证明的一个方案；第二个黑体句描述了一个判断，用来支持那个方案的证明。第一个黑体句描述的方案是论证不支持的。

(B) 第一个黑体句描述了论证试图去解释的一个方案；第二个黑体句给出了论证对于这个方案的部分解释。第一个黑体句描述的方案是论证不支持的。

(C) 第一个黑体句描述了论证试图去评估的一个方案；第二个黑体句认为这个方案不能达成它的目标。第二个黑体句并不是说现有方案不能达成目标，而是说有替代方案可以备选。

(D) 正确。第一个黑体句描述了论证试图去评估的一个方案；第二个黑体句认为这个方案是不需要的。

(E) 第一个黑体句对论证试图去评估的方案提供了支持；第二个黑体句提供了这个方案应该被废除的证据。第一个黑体句本身就是一个方案，不是对方案的支持。

例题 3

Sasha: It must be healthy to follow a diet high in animal proteins and fats. Human beings undoubtedly evolved to thrive on such a diet, since our prehistoric ancestors ate large amounts of meat.

Jamal: But our ancestors also exerted themselves intensely in order to obtain this food, whereas most human beings today are much less physically active.

Jamal responds to Sasha by doing which of the following?

(A) Refuting her statement about our prehistoric ancestors

(B) Bringing forth a piece of information for the purpose of suggesting that she should qualify her main conclusion

(C) Citing additional evidence that indirectly supports her conclusion and suggests a way to broaden it

(D) Questioning whether her assumption about our prehistoric ancestors permits any conclusions about human evolution

(E) Expressing doubts about whether most human beings today are as healthy as our prehistoric ancestors were

萨莎的陈述：萨莎认为富含动物蛋白和脂肪的饮食是健康的。她的理由是，人类进化出了适应这种饮食的方式，因为我们的史前祖先吃了大量的肉。

贾迈尔的回应：贾迈尔提出了一个观点，即我们的祖先在获取这种食物时进行了大量的体力活动，而今天大多数人的体力活动量则少得多。

选项分析

(A) 反驳她的关于我们史前祖先的陈述。贾迈尔并没有直接反驳萨莎的关于史前祖先饮食的说法。相反，他补充了史前祖先生活方式的相关信息。

(B) 正确。贾迈尔提出一个信息，目的是建议她对主要结论进行限定。这与贾迈尔的回应最为贴近。他提出了对体力活动的考量，这表明萨莎关于饮食的结论在现代生活方式的背景下，可能需要重新评估或限定。

(C) 引用间接支持她结论的额外证据，并建议拓宽她的结论。贾迈尔的回应并不支持萨莎的结论。相反，它引入了一个不同的方面（体力活动），这影响了饮食的整体健康性。

(D) 质疑她的关于我们史前祖先的假设是否允许对人类进化做出任何结论。贾迈尔并没有直接质疑萨莎的关于人类进化或我们祖先饮食的假设。他的重点更多关注现在和过去的人的体力活动的差异。

(E) 对今天大多数人是否和我们史前祖先一样健康表示怀疑。尽管贾迈尔暗示了体力活动的差异导致了健康差异，但他并没有直接比较现代人与我们史前祖先的健康状况。

例题 4

The Sumpton town council recently voted to pay a prominent artist to create an abstract sculpture for the town square. Critics of this decision protested that town residents tend to dislike most abstract art, and any art in the town square should reflect their tastes. But a town council spokesperson dismissed this criticism, pointing out that other public abstract sculptures that the same sculptor has installed in other cities have been extremely popular with those cities' local residents.

The statements above most strongly suggest that the main point of disagreement between the critics and the spokesperson is whether

(A) it would have been reasonable to consult town residents on the decision

(B) most Sumpton residents will find the new sculpture to their tastes

(C) abstract sculptures by the same sculptor have truly been popular in other cities

(D) a more traditional sculpture in the town square would be popular among local residents

(E) public art that the residents of Sumpton would find desirable would probably be found desirable by the residents of other cities

这个问题涉及苏普顿镇议会对于在镇中心安装一位著名艺术家的抽象雕塑的决定，以及对此决定的批评和回应。让我们逐一分析每个选项，以确定批评者和发言人之间的主要分歧点。

选项分析

(A) 咨询镇上居民对于这个决定的意见是否合理。批评者认为镇上的艺术品应该反映居民的品位，这似乎暗示了应该咨询居民。但议会发言人的回应并未直接提及咨询居民的问题，而是指出这位艺术家在其他城市创作的抽象雕塑非常受欢迎。

(B) 正确。大多数苏普顿镇居民是否会喜欢这个新雕塑。这是批评者和发言人之间的主要分歧。批评者认为大多数居民不喜欢抽象艺术，而发言人则暗示居民会喜欢这位艺术家的作品。

(C) 同一位雕塑家的抽象雕塑在其他城市是否真的受欢迎。发言人的论点基于这位雕塑家在其他城市的作品受欢迎这一事实，但这并不是双方分歧的核心。

(D) 镇中心如果放置更传统的雕塑是否会受到当地居民的欢迎。这个观点在双方的陈述中没有直接提到，虽然它与整体讨论有关，但并不是主要的分歧点。

(E) 苏普顿镇居民会喜欢的公共艺术品是否也会被其他城市的居民所喜欢。这个观点同样没有在双方的陈述中直接讨论。

例题 5

Most scholars agree that King Alfred (A. D. 849—899) personally translated a number of Latin texts into Old English. One historian contends that Alfred also personally penned his own law code, arguing that the numerous differences between the language of the law code and Alfred's translations of Latin texts are outweighed by the even more numerous similarities. Linguistic similarities, however, are what one expects in texts from the same language, the same time, and the same region. Apart from Alfred's surviving translations and law code, there are only two other extant works from the same dialect and milieu, so it is risky to assume here that linguistic similarities point to common authorship.

The passage above proceeds by

(A) providing examples that underscore another argument's conclusion
(B) questioning the plausibility of an assumption on which another argument depends
(C) showing that a principle if generally applied would have anomalous consequences
(D) showing that the premises of another argument are mutually inconsistent
(E) using argument by analogy to undermine a principle implicit in another argument

这段文字讨论了有关英国国王阿尔弗雷德（公元 849—899 年）是否亲自撰写了法典的历史学问题。让我们分析每个选项，并确定这段文字的论证方式。

选项分析

(A) 提供例子来强调另一个论点的结论。这段文字没有提供具体例子来强调或支持其他论点的结论。

(B) 正确。质疑另一个论点所依赖的假设的合理性。这段文字确实在质疑一个假设的合理性，即基于语言相似性推断共同作者的假设。历史学家认为法典和阿尔弗雷德翻译的拉丁文本之间的语言相似性表明它们有共同的作者。而这段文字指出，由于只有很少的作品来自同一地区、使用同一种语言，所以依赖语言相似性来推断作者是有风险的。

(C) 表明如果普遍应用一个原则会导致异常后果。这段文字并没有讨论普遍应用某个原则会带来的后果。

(D) 表明另一个论点的前提相互矛盾。这里并没有指出任何论点的前提是相互矛盾的。

(E) 使用类比论证来削弱另一个论点中隐含的原则。这段文字没有使用类比来论证或削弱任何隐含的原则。

练习

1. One of the limiting factors in human physical performance is the amount of oxygen that is absorbed by the muscles from the bloodstream. Accordingly, entrepreneurs have begun selling at gymnasiums and health clubs bottles of drinking water, labeled "SuperOXY" that has extra oxygen dissolved in the water. Such water would be useless in improving physical performance, however, since **the amount of oxygen in the blood of someone who is exercising is already more than the muscle cells can absorb**.

Which of the following, if true, would serve the same function in the argument as the statement in boldface?

(A) World-class athletes turn in record performances without such water.

(B) Frequent physical exercise increases the body's ability to take in and use oxygen.

(C) The only way to get oxygen into the bloodstream so that it can be absorbed by the muscles is through the lungs.

(D) Lack of oxygen is not the only factor limiting human physical performance.

(E) The water lost in exercising can be replaced with ordinary tap water.

2. In countries where automobile insurance includes compensation for whiplash injuries sustained in automobile accidents, reports of having suffered such injuries are twice as frequent as they are in countries where whiplash is not covered. Some commentators have argued, correctly, that since **there is presently no objective test for whiplash, spurious reports of whiplash injuries cannot be readily identified**. These commentators are, however, wrong to draw the further conclusion that **in the countries with the higher rates of reported whiplash injuries, half of the reported cases are spurious**: clearly, in countries where automobile insurance does not include compensation for whiplash,

people often have little incentive to report whiplash injuries that they actually have suffered.

In the argument given, the two boldfaced portions play which of the following roles?

(A) The first is evidence that has been used to support a conclusion that the argument criticizes; the second is that conclusion.
(B) The first is evidence that has been used to support a conclusion that the argument criticizes; the second is the position that the argument defends.
(C) The first is a claim that has been used to support a conclusion that the argument accepts; the second is the position that the argument defends.
(D) The first is an intermediate conclusion that has been used to support a conclusion that the argument defends; the second is the position that the argument opposes.
(E) The first presents a claim that is disputed in the argument; the second is a conclusion that has been drawn on the basis of that claim.

3. Most of Western music since the Renaissance has been based on a seven-note scale known as the diatonic scale, but when did the scale originate? **A fragment of a bone flute excavated at a Neanderthal campsite has four holes, which are spaced in exactly the right way for playing the third through sixth notes of a diatonic scale.** The entire flute must surely have had more holes, and the flute was made from a bone that was long enough for these additional holes to have allowed a complete diatonic scale to be played. **Therefore, the Neanderthals who made the flute probably used a diatonic musical scale.**

In the argument given, the two portions in boldface play which of the following roles?

(A) The first introduces evidence to support the main conclusion of the argument; the second is the main conclusion stated in the argument.

(B) The first introduces evidence to support the main conclusion of the argument; the second presents a position to which the argument is opposed.

(C) The first describes a discovery as undermining the position against which the argument as a whole is directed; the second states the main conclusion of the argument.

(D) The first introduces the phenomenon that the argument as a whole seeks to explain; the second presents a position to which the argument is opposed.

(E) The first introduces the phenomenon that the argument as a whole seeks to explain; the second gives a reason to rule out one possible explanation.

4. Journalist: **Every election year at this time the state government releases the financial disclosures that potential candidates must make in order to be eligible to run for office.** Among those making the required financial disclosure this year is a prominent local businessman, Arnold Bergeron. There has often been talk in the past of Mr. Bergeron's running for governor, not least from Mr. Bergeron himself. **This year it is likely that he finally will, since those who have discounted the possibility of a Bergeron candidacy have always pointed to the necessity of making financial disclosure as the main obstacle to such a candidacy.**

In the journalist's argument, the two boldfaced portions play which of the following roles?

(A) The first provides information without which the argument lacks force; the second states the main conclusion of the argument.

(B) The first provides information without which the argument lacks force; the second states an intermediate conclusion that is used to support a further conclusion.

(C) The first describes a practice that the journalist seeks to defend; the second cites a likely consequence of this practice.

(D) The first states evidence bearing against the main conclusion of the argument; the second states that conclusion.

(E) Each provides evidence in support of an intermediate conclusion that supports a further conclusion stated in the argument.

5. A prominent investor who holds a large stake in the Burton Tool Company has recently claimed that the company is mismanaged. As evidence for this claim, the investor cited the company's failure to slow production in response to a recent rise in its inventory of finished products. It is doubtful whether an investor's sniping at management can ever be anything other than counterproductive, but **in this case it is clearly not justified**. It is true that an increased inventory of finished products often indicates that production is outstripping demand. **In Burton's case it indicates no such thing**, however, the increase in inventory is entirely attributable to products that have already been assigned to orders received from customers.

In the argument given, the two boldfaced portions play which of the following roles?

(A) The first provides evidence to support the conclusion of the argument as a whole; the second states that conclusion.

(B) The first states the conclusion of the argument as a whole; the second states an intermediate conclusion that is drawn in order to support that conclusion.

(C) The first is the position that the argument as a whole opposes; the second provides evidence against the position being opposed.

(D) The first states an intermediate conclusion that is drawn in order to support the conclusion of the argument as a whole; the second states the conclusion of the argument as a whole.

(E) The first and the second both state intermediate conclusions that are drawn in order to support jointly the conclusion of the argument as a whole.

6. Last year a record number of new manufacturing jobs were created. Will this year bring another record? Well, **any new manufacturing job is created either within an existing company or by the start-up of a new company**. Within existing firms, new jobs have been created this year at well below last year's record pace. At the same time, there is considerable evidence that the number of new companies starting up will be no higher this year than it was last year and surely **the new companies starting up this year will create no more jobs per company than did last year's start-ups**. So clearly, the number of new jobs created this year will fall short of last year's record.

In the argument given, the two portions in boldface play which of the following roles?

(A) The first is presented as an obvious truth on which the argument is based; the second is a prediction advanced in support of the main conclusion of the argument.
(B) The first is presented as an obvious truth on which the argument is based; the second is an objection that the argument rejects.
(C) The first is presented as an obvious truth on which the argument is based; the second is the main conclusion of the argument.
(D) The first is a generalization that the argument seeks to establish; the second is a claim that has been advanced in support of a position that the argument opposes.
(E) The first is a generalization that the argument seeks to establish; the second is a claim that has been advanced in order to challenge that generalization.

7. **Environmental organizations want to preserve the land surrounding the Wilgrinn Wilderness Area from residential development.** They plan to do this by purchasing that land from the farmers who own it. That plan is ill-conceived: if the farmers did sell their land, they would sell it to the highest bidder, and developers would outbid any other bidders. On the other hand, **these farmers will never actually sell any of the land, provided that farming it remains viable**. But farming will not remain viable if the farms are left unmodernized, and most of the farmers lack the financial resources modernization requires. And that is exactly why a more sensible preservation strategy would be to assist the farmers to modernize their farms to the extent needed to maintain viability.

In the argument as a whole, the two boldface proportions play which of the following roles?

(A) The first presents a goal that the argument rejects as ill-conceived; the second is evidence that is presented as grounds for that rejection.

(B) The first presents a goal that the argument concludes cannot be attained; the second is a reason offered in support of that conclusion.

(C) The first presents a goal that the argument concludes can be attained; the second is a judgment disputing that conclusion.

(D) The first presents a goal, strategies for achieving which are being evaluated in the argument; the second is a judgment providing a basis for the argument's advocacy of a particular strategy.

(E) The first presents a goal that the argument endorses; the second presents a situation that the argument contends must be changed if that goal is to be met in the foreseeable future.

8. Shirla: In figure skating competitions that allow amateur and professional skaters to compete against each other, the professionals are bound to have an unfair advantage. After all, most of them became professional only after success on the amateur circuit.

Ron: But that means that it's been a long time since they've had to meet the more rigorous technical standards of the amateur circuit.

Which of the following is most likely a point at issue between Shirla and Ron?

(A) Whether there should be figure skating competitions that allow amateur and professional skaters to compete against each other

(B) Whether the scores of professional skaters competing against amateurs should be subject to adjustment to reflect the special advantages of professionals

(C) Whether figure skaters can successfully become professional before success on the amateur circuit

(D) Whether the technical standards for professional figure skating competition are higher than those for amateur figure skating competition

(E) Whether professional figure skaters have an unfair advantage over amateur figure skaters in competitions in which they compete against each other

9. Mansour: We should both plan to change some of our investments from coal companies to less polluting energy companies. And here's why. Consumers are increasingly demanding nonpolluting energy, and energy companies are increasingly supplying it.

Therese: I'm not sure we should do what you suggest. As demand for nonpolluting energy increases relative to supply, its price will increase, and then the more polluting energy will cost relatively less. Demand for the cheaper, dirtier energy forms will then increase, as will the stock values of the companies that produce them.

Therese responds to Mansour's proposal by doing which of the following?

(A) Advocating that consumers use less expensive forms of energy

(B) Implying that not all uses of coal for energy are necessarily polluting

(C) Disagreeing with Mansour's claim that consumers are increasingly demanding nonpolluting energy

(D) Suggesting that leaving their existing energy investments unchanged could be the better course

(E) Providing a reason to doubt Mansour's assumption that supply of nonpolluting energy will increase in line with demand

10. Environmentalist: The use of snowmobiles in the vast park north of Milville creates unacceptable levels of air pollution and should be banned.

Milville business spokesperson: Snowmobiling brings many out-of-towners to Milville in winter months, to the great financial benefit of many local residents. So, economics dictate that we put up with the pollution.

Environmentalist: I disagree: A great many cross-country skiers are now kept from visiting Milville by the noise and pollution that snowmobiles generate.

Environmentalist responds to the business spokesperson by doing which of the following?

(A) Challenging an assumption that certain desirable outcome can derive from only one set of circumstances.

(B) Challenging an assumption that certain desirable outcome is outweighed by negative aspects associated with producing that outcome.

(C) Maintaining that the benefit that the spokesperson desires could be achieved in greater degree by a different means.

(D) Claiming that the spokesperson is deliberately misrepresenting the environmentalist's position in order to be better able to attack it.

(E) Denying that an effect that the spokesperson presents as having benefited a certain group of people actually benefited those people.

答案及解析

1. One of the limiting factors in human physical performance is the amount of oxygen that is absorbed by the muscles from the bloodstream. Accordingly, entrepreneurs have begun selling at gymnasiums and health clubs bottles of drinking water, labeled "SuperOXY" that has extra oxygen dissolved in the water. Such water would be useless in improving physical performance, however, since **the amount of oxygen in the blood of someone who is exercising is already more than the muscle cells can absorb**.

Which of the following, if true, would serve the same function in the argument as the statement in boldface?

(A) World-class athletes turn in record performances without such water.

(B) Frequent physical exercise increases the body's ability to take in and use oxygen.

(C) The only way to get oxygen into the bloodstream so that it can be absorbed by the muscles is through the lungs.

(D) Lack of oxygen is not the only factor limiting human physical performance.

(E) The water lost in exercising can be replaced with ordinary tap water.

本题问法比较特殊，要求我们从五个选项中找到一个与黑体句作用相同的选项。实际上，通过分析推理文段可知，黑体句是主结论"Such water would be useless in improving physical performance."的原因，因此我们去找"Such water would be useless in improving physical performance."的另外一个原因即可。

选项分析：

(A) 世界级运动员不喝这样的水也能创纪录。本选项不能解释为什么这些水没用，因为运动员不喝这些水也可以创纪录，不代表喝了没用。

(B) 高频率的体育运动可以增强身体摄入和用氧的能力。这是人体吸收氧的原理，不能说明结论的产生。

(C) 正确。人体唯一能吸收氧并让其进入血液而最终被肌肉吸收的方式就是通过肺。显然，本选项可以解释为什么喝此类含氧水没用。

(D) 缺氧并非影响人们运动能力的唯一因素。即使有别的因素影响运动能力，我们要解释的是“缺氧”这个因素为什么不能通过喝含氧水解决。

(E) 在运动中缺失的水可以直接用普通自来水来弥补。本选项是说明“补充”水的原理，不能说明为什么 OXY 水没有用。

2. In countries where automobile insurance includes compensation for whiplash injuries sustained in automobile accidents, reports of having suffered such injuries are twice as frequent as they are in countries where whiplash is not covered. Some commentators have argued, correctly, that since **there is presently no objective test for whiplash, spurious reports of whiplash injuries cannot be readily identified**. These commentators are, however, wrong to draw the further conclusion that **in the countries with the higher rates of reported whiplash injuries, half of the reported cases are spurious**: clearly, in countries where automobile insurance does not include compensation for whiplash, people often have little incentive to report whiplash injuries that they actually have suffered.

In the argument given, the two boldfaced portions play which of the following roles?

(A) The first is evidence that has been used to support a conclusion that the argument criticizes; the second is that conclusion.

(B) The first is evidence that has been used to support a conclusion that the argument criticizes; the second is the position that the argument defends.

(C) The first is a claim that has been used to support a conclusion that the argument accepts; the second is the position that the argument defends.

(D) The first is an intermediate conclusion that has been used to support a conclusion that the argument defends; the second is the position that the argument opposes.

(E) The first presents a claim that is disputed in the argument; the second is a conclusion that has been drawn on the basis of that claim.

确定文段的主结论并判断黑体句与主结论的关系。

本文段的主结论为：These commentators are, however, wrong to draw the further conclusion.

第一个黑体句是commentators得出结论的依据，而主结论在反对commentators得出的结论，所以第一个黑体句与主结论相矛盾；第二个黑体句就是主结论不同意的那个结论，也与主结论相矛盾。

选项分析：

(A) 正确。第一个黑体句是一个证据，用来支持文段所批判的结论；第二个黑体句是那个结论。

(B) 第一个黑体句是一个证据，用来支持文段所批判的结论；第二个黑体句是文段维护的立场。第二个黑体句的描述有误，论证最后是不同意这个结论的，并没有维护这个结论。

(C) 第一个黑体句是一个声称，用来支持文段所接受的结论；第二个黑体句是文段的立场。两句的描述均有误，第一句没有支持主结论，第二句也不是主结论。

(D) 第一个黑体句是一个中间结论，用来支持文段所持有的结论；第二个黑体句是文段所反驳的立场。第一个黑体句是一个证据而不是中间结论。

(E) 第一个黑体句是一个声称，在文段中被质疑；第二个黑体句是基于这个声明而产生的结论。文段并没有对第一个黑体句所描述的内容产生任何异议。

3. Most of Western music since the Renaissance has been based on a seven-note scale known as the diatonic scale, but when did the scale originate? **A fragment of a bone flute excavated at a Neanderthal campsite has four holes, which are spaced in exactly the right way for playing the third through sixth notes of a diatonic scale.** The entire flute must surely have had more holes, and the flute was made from a bone that was long enough for these additional holes to have allowed a complete diatonic scale to be played. **Therefore, the Neanderthals who made the flute probably used a diatonic musical scale.**

In the argument given, the two portions in boldface play which of the following roles?

(A) The first introduces evidence to support the main conclusion of the argument; the second is the main conclusion stated in the argument.

(B) The first introduces evidence to support the main conclusion of the argument; the second presents a position to which the argument is opposed.

(C) The first describes a discovery as undermining the position against which the argument as a whole is directed; the second states the main conclusion of the argument.

(D) The first introduces the phenomenon that the argument as a whole seeks to explain; the second presents a position to which the argument is opposed.

(E) The first introduces the phenomenon that the argument as a whole seeks to explain; the second gives a reason to rule out one possible explanation.

确定文段的主结论并判断黑体部分与主结论的关系。

本文段的主结论为：The Neanderthals who made the flute probably used a diatonic musical scale.

第一个黑体句是描述了一个现象，这个现象在论证的最后探究了原因。第二个黑体句给出了这个现象产生的原因，也就是整个文段的主结论。

选项分析：

(A) 正确。第一个黑体句介绍了支持主结论的证据；第二个黑体句是文段的主结论。

(B) 第一个黑体句介绍了支持论证主结论的证据；第二个黑体句表达了文段反驳的一个立场。第二个黑体句所表达的立场是文段同意的立场。

(C) 第一个黑体句将一个发现描述为削弱整个文段所针对的立场；第二个黑体句是论证的主结论。第一个黑体句是第二个黑体句的前提，也就是论证主结论的前提。

(D) 第一个黑体句描述了文段试图去解释的一个现象；第二个黑体句表达了文段反驳的立场。第二个黑体句所表达的立场是文段同意的立场。

(E) 第一个黑体句描述了文段试图去解释的一个现象；第二个黑体句给出了删除一个可能原因的理由。第二个黑体句所表达的立场是文段同意的立场。

4. Journalist: **Every election year at this time the state government releases the financial disclosures that potential candidates must make in order to be eligible to run for office.** Among those making the required financial disclosure this year is a prominent local businessman, Arnold Bergeron. There has often been talk in the past of Mr. Bergeron's running for governor, not least from Mr. Bergeron himself. **This year it is likely that he finally will, since those who have discounted the possibility of a Bergeron candidacy have always pointed to the necessity of making financial disclosure as the main obstacle to such a candidacy.**

In the journalist's argument, the two boldfaced portions play which of the following roles?

(A) The first provides information without which the argument lacks force; the second states the main conclusion of the argument.

(B) The first provides information without which the argument lacks force; the second states an intermediate conclusion that is used to support a further conclusion.

(C) The first describes a practice that the journalist seeks to defend; the second cites a likely consequence of this practice.

(D) The first states evidence bearing against the main conclusion of the argument; the second states that conclusion.

(E) Each provides evidence in support of an intermediate conclusion that supports a further conclusion stated in the argument.

确定文段的主结论并判断黑体部分与主结论的关系。

本文段的主结论为：This year it is likely that he finally will.

第一个黑体句是前提，第二个黑体句是结论。前提的作用以及目的只有一个，即支持结论。

选项分析：

(A) 正确。第一个黑体句提供了信息，没有这些信息，论证就缺乏说服力；第二个黑体句声明了文章的主结论。

(B) 第一个黑体句提供了信息，没有这些信息，论证就缺乏说服力；第二个黑体句给出了一个用于支持进一步结论的中间结论。第二个黑体句不是中间结论，而是最后的结论。

(C) 第一个黑体句描述了记者想要维护的一个实践；第二个黑体句给出了这个实践的可能结果。第一个黑体句并不是描述记者想要维护的一个实践，而是一个中性的信息。

(D) 第一个黑体句阐述了一个反对主结论的证据；第二个黑体句是论证的主结论。第一个黑体句是论证最后结论的前提，并没有反对主结论。

(E) 每一个黑体句都提供了支持中间结论的证据，而中间结论又支持文段进一步的结论。

5. A prominent investor who holds a large stake in the Burton Tool Company has recently claimed that the company is mismanaged. As evidence for this claim, the investor cited the company's failure to slow production in response to a recent rise in its inventory of finished products. It is doubtful whether an investor's sniping at management can ever be anything other than counterproductive, but **in this case it is clearly not justified**. It is true that an increased inventory of finished products often indicates that production is outstripping demand. **In Burton's case it indicates no such thing**, however, the increase in inventory is entirely attributable to products that have already been assigned to orders received from customers.

In the argument given, the two boldfaced portions play which of the following roles?

(A) The first provides evidence to support the conclusion of the argument as a whole; the second states that conclusion.

(B) The first states the conclusion of the argument as a whole; the second states an intermediate conclusion that is drawn in order to support that conclusion.

(C) The first is the position that the argument as a whole opposes; the second provides evidence against the position being opposed.

(D) The first states an intermediate conclusion that is drawn in order to support the conclusion of the argument as a whole; the second states the conclusion of the argument as a whole.

(E) The first and the second both state intermediate conclusions that are drawn in order to support jointly the conclusion of the argument as a whole.

确定文段的主结论并判断黑体句与主结论的关系。

本文段的主结论为：In this case it is clearly not justified.

第一个黑体句是主结论。主结论之后的内容全是在论证主结论，所以第二个黑体句是支持主结论的。

选项分析：

(A) 第一个黑体句提供了一个证据来支持主结论；第二个黑体句给出了这个结论。第一个黑体句是文段的主结论。

(B) 正确。第一个黑体句是文段的主结论；第二个黑体句给出了一个支持这个主结论的中间结论。

(C) 第一个黑体句是文段所反驳的立场；第二个黑体句给出了一个反对文段立场的证据。第一个黑体句是主结论。

(D) 第一个黑体句是一个支持主结论的中间结论；第二个黑体句是主结论。第一个黑体句才是主结论。

(E) 第一个和第二个黑体句均是支持主结论的中间结论。第一个黑体句是论证的主结论。

6. Last year a record number of new manufacturing jobs were created. Will this year bring another record? Well, **any new manufacturing job is created either**

within an existing company or by the start-up of a new company. Within existing firms, new jobs have been created this year at well below last year's record pace. At the same time, there is considerable evidence that the number of new companies starting up will be no higher this year than it was last year and surely **the new companies starting up this year will create no more jobs per company than did last year's start-ups**. So clearly, the number of new jobs created this year will fall short of last year's record.

In the argument given, the two portions in boldface play which of the following roles?

(A) The first is presented as an obvious truth on which the argument is based; the second is a prediction advanced in support of the main conclusion of the argument.

(B) The first is presented as an obvious truth on which the argument is based; the second is an objection that the argument rejects.

(C) The first is presented as an obvious truth on which the argument is based; the second is the main conclusion of the argument.

(D) The first is a generalization that the argument seeks to establish; the second is a claim that has been advanced in support of a position that the argument opposes.

(E) The first is a generalization that the argument seeks to establish; the second is a claim that has been advanced in order to challenge that generalization.

确定文段的主结论并判断黑体句与主结论的关系。

本文段的主结论为：The number of new jobs created this year will fall short of last year's record.

这个结论句之外的句子，我们都可以认为是这个结论句的前提。所以，两个黑体句均是支持主结论的。

选项分析：

(A) 正确。第一个黑体句给出了文段所基于的一个明显的真相；第二个黑体句是支持主结论的一个前提。

(B) 第一个黑体句给出了文段所基于的一个明显的真相；第二个黑体句是文段所反驳的反对意见。第二个黑体句是支持主结论的一个前提。

(C) 第一个黑体句给出了文段所基于的一个明显的真相；第二个黑体句是主结论。第二个黑体句是支持主结论的一个前提。

(D) 第一个黑体句是文段试图建立的一个概论；第二个黑体句是一个声明，用来支持文段所反对的结论。第二个黑体句是支持主结论的一个前提。

(E) 第一个黑体句是文段试图建立的一个概论；第二个黑体句是反驳这个概论的一个声明。第二个黑体句是支持主结论的一个前提。

7. **Environmental organizations want to preserve the land surrounding the Wilgrinn Wilderness Area from residential development.** They plan to do this by purchasing that land from the farmers who own it. That plan is ill-conceived: if the farmers did sell their land, they would sell it to the highest bidder, and developers would outbid any other bidders. On the other hand, **these farmers will never actually sell any of the land, provided that farming it remains viable**. But farming will not remain viable if the farms are left unmodernized, and most of the farmers lack the financial resources modernization requires. And that is exactly why a more sensible preservation strategy would be to assist the farmers to modernize their farms to the extent needed to maintain viability.

In the argument as a whole, the two boldface proportions play which of the following roles?

(A) The first presents a goal that the argument rejects as ill-conceived; the second is evidence that is presented as grounds for that rejection.

(B) The first presents a goal that the argument concludes cannot be attained; the second is a reason offered in support of that conclusion.

(C) The first presents a goal that the argument concludes can be attained; the second is a judgment disputing that conclusion.

(D) The first presents a goal, strategies for achieving which are being evaluated in the argument; the second is a judgment providing a basis for the argument's advocacy of a particular strategy.

(E) The first presents a goal that the argument endorses; the second presents a situation that the argument contends must be changed if that goal is to be met in the foreseeable future.

确定文段的主结论并判断黑体句与主结论的关系。

本文段的主结论为：And that is exactly why a more sensible preservation strategy would be to assist the farmers to modernize their farms to the extent needed to maintain viability.

第一个黑体句是整个文段的背景，即环保组织想要达成的目标。后面的内容评估了为了实现这个目标而打算采取的方案。第二个黑体句反对了第一个方案，同时也支持了主结论中陈述的第二个方案。

选项分析：

(A) 第一个黑体句给出了文段反对的、认为有问题的目标；第二个黑体句支持这个反对。文段没有说这个目标有问题。
(B) 第一个黑体句给出了文段认为无法实现的目标；第二个黑体句是支持这个结论的理由。文段没有说这个目标不能达到。
(C) 第一个黑体句给出了文段认为无法实现的目标；第二个黑体句是反对这个结论的评价。文段没有说这个目标不能达到。
(D) 正确。第一个黑体句给出了一个目标，用来实现此目标的策略在文段中被评估过；第二个黑体句为文段支持那个策略提供基础。
(E) 第一个黑体句给出了文段支持的目标；第二个黑体句给出了一个场景，文段认为如果想要在可预见的未来实现这一目标，必须要改变此场景。文章没有企图改变农民的想法，而是要改变我们的策略，去顺应农民。

8. Shirla：In figure skating competitions that allow amateur and professional skaters to compete against each other, the professionals are bound to have an unfair advantage. After all, most of them became professional only after success on the amateur circuit.

Ron：But that means that it's been a long time since they've had to meet the more rigorous technical standards of the amateur circuit.

Which of the following is most likely a point at issue between Shirla and Ron?

(A) Whether there should be figure skating competitions that allow amateur and professional skaters to compete against each other

(B) Whether the scores of professional skaters competing against amateurs should be subject to adjustment to reflect the special advantages of professionals

(C) Whether figure skaters can successfully become professional before success on the amateur circuit

(D) Whether the technical standards for professional figure skating competition are higher than those for amateur figure skating competition

(E) Whether professional figure skaters have an unfair advantage over amateur figure skaters in competitions in which they compete against each other

希拉的陈述：在花样滑冰比赛中，业余选手和专业选手相互竞争，专业选手必然会拥有不公平的优势。毕竟，他们中的大多数人都是在业余巡回赛上取得成功后才成为职业选手的。

罗恩的回应：但这意味着他们已经很久没有达到业余巡回赛更严格的技术标准了。

问题问希拉和罗恩的争论点是什么。只需找出两人各自的结论即可。

希拉的结论是：The professionals are bound to have an unfair advantage.

罗恩虽然没有明确说出自己的结论，但通过 but 一词可以看出，他不赞成希拉的观点。

所以，两人的争论点就在于“专业人士究竟有没有不公平的优势”。

选项分析：

(A) 是否应该举行允许业余和专业滑冰选手相互竞争的花样滑冰比赛。两人争论的是专业选手是否拥有不公平的优势，至于后续能否举行允许业余和专业选手相互竞争的比赛，并不在目前的讨论范围内。

(B) 专业选手与业余选手的比赛成绩是否应进行调整，以体现专业选手的特殊优势。两人争论的是专业选手是否拥有不公平的优势，至于后续是否要对比赛成绩进行调整，并不在目前的讨论范围内。

(C) 花样滑冰运动员在业余巡回赛上取得成功之前能否成功地成为专业选手。文段完全没提及。

(D) 花样滑冰专业比赛的技术水平是否高于花样滑冰业余比赛的技术水平。两人并没有就技术水平展开争论。

(E) 正确。专业花样滑冰运动员在相互竞争的比赛中是否比业余花样滑冰运动员拥有不公平的优势。

9. Mansour: We should both plan to change some of our investments from coal companies to less polluting energy companies. And here's why. Consumers are increasingly demanding nonpolluting energy, and energy companies are increasingly supplying it.

Therese: I'm not sure we should do what you suggest. As demand for nonpolluting energy increases relative to supply, its price will increase, and then the more polluting energy will cost relatively less. Demand for the cheaper, dirtier energy forms will then increase, as will the stock values of the companies that produce them.

Therese responds to Mansour's proposal by doing which of the following?

(A) Advocating that consumers use less expensive forms of energy

(B) Implying that not all uses of coal for energy are necessarily polluting

(C) Disagreeing with Mansour's claim that consumers are increasingly demanding nonpolluting energy

(D) Suggesting that leaving their existing energy investments unchanged could be the better course

(E) Providing a reason to doubt Mansour's assumption that supply of nonpolluting energy will increase in line with demand

曼苏尔的陈述：我们都应该计划将一些投资从煤炭公司转向污染更少的能源公司。

原因如下：消费者对无污染能源的要求越来越高，能源公司也越来越多地提供这种能源。

特蕾莎的回应：我不确定我们是否应该按照你的建议去做。当对无污染能源的需求相对于供给增加时，其价格会上升，相对而言，污染较大的能源成本相对较低。那么，对更便宜、更脏的能源的需求将会增加，生产这些能源的公司的股价也会上涨。

问题问特蕾莎对曼苏尔的建议做出了什么回应。只需找出两人各自的结论即可。

曼苏尔的主结论是：我们都应该计划将一些投资从煤炭公司转向污染更少的能源公司。

特蕾莎的主结论是：我不确定我们是否应该按照你的建议去做。

选项分析：

(A) 提倡消费者使用更便宜的能源。特蕾莎并没有倡导消费者应该做什么。

(B) 表明并非所有的煤炭能源使用都必然造成污染。特蕾莎没有讨论煤炭能源造成污染的情况。

(C) 不同意曼苏尔关于消费者越来越需要无污染能源的说法。特蕾莎只是建议不要轻易改变投资方向，并没反对“消费者越来越不需要无污染能源”这个观点。消费者在需要无污染能源的基础上，有可能也不会减少对污染能源的需求。

(D) 正确。表明他们保持现有的能源投资不变可能是更好的做法。

(E) 提供了一个理由来怀疑曼苏尔的假设，即无污染能源的供应将随着需求的增加而增加。特蕾莎没有针对无污染能源的供给情况做任何反驳。

10. Environmentalist: The use of snowmobiles in the vast park north of Milville creates unacceptable levels of air pollution and should be banned.

Milville business spokesperson: Snowmobiling brings many out-of-towners to Milville in winter months, to the great financial benefit of many local residents.

So, economics dictate that we put up with the pollution.

Environmentalist: I disagree: A great many cross-country skiers are now kept from visiting Milville by the noise and pollution that snowmobiles generate.

Environmentalist responds to the business spokesperson by doing which of the following?

(A) Challenging an assumption that certain desirable outcome can derive from only one set of circumstances.

(B) Challenging an assumption that certain desirable outcome is outweighed by negative aspects associated with producing that outcome.

(C) Maintaining that the benefit that the spokesperson desires could be achieved in greater degree by a different means.

(D) Claiming that the spokesperson is deliberately misrepresenting the environmentalist's position in order to be better able to attack it.

(E) Denying that an effect that the spokesperson presents as having benefited a certain group of people actually benefited those people.

环保主义者的陈述：在米尔维尔北部广阔的公园里使用雪地摩托造成了不可接受的空气污染，应该禁止使用雪地摩托。

米尔维尔商业发言人的回应：雪地摩托在冬季给米尔维尔带来了许多外地人，给许多当地居民带来了巨大的经济利益。所以，经济要求我们忍受污染。

环保主义者进一步回应：我不同意。由于雪地摩托产生的噪音和污染，很多越野滑雪者现在都不去米尔维尔了。

选项分析：

(A) 质疑一种假设，即某种理想结果只能从一组环境中产生。环保主义者没有讲理想结果的来源有几种环境。

(B) 质疑一种假设，即某种理想结果被与产生该结果相关的负面因素所压倒。环保主义者应该是支持“负面结果超过了理想结果”的。

(C) 认为发言人所期望的利益可以通过不同的方式在更大程度上实现。环保主义者没有提出多种方式。

(D) 声称发言人故意歪曲环保主义者的立场，以便更好地攻击它。环保主义者并没有认为发言人在故意歪曲自己的立场。

(E) 正确。否认发言人所说的有利于某一群体的结果实际上使那些人受益。发言人认为会有游客被吸引过来，从而给当地人带来收益。但环保主义者认为这些人实际上会被污染劝退，从而没法给当地人带来收益。

2.2 构建论证（Construction）

构建论证类考题主要分为“确定结论”“确定前提”和“现象解释”三类。

2.2.1 确定结论

顾名思义，这类考题会给出我们一些已知信息，要求我们运用“演绎推理”的能力，选出这些信息蕴含的结论。例如：

小明每周三一定会穿衬衫。今天是周三。

如果上述说法正确，那么以下哪个说法一定正确？

(A) 小明今天会穿衬衫。

(B) 小明喜欢穿衬衫。

例题答案为选项 A 。这类考题的本质是在问题干信息“蕴含”下列哪个选项。显然，选项 B 是不一定会发生的，即我们无法确定小明是否喜欢穿衬衫。

GMAT 考试也会借助模态命题、逻辑运算符和量词（详见 1. 1 演绎论证）来提升考题难度。例如：

如果一个动物是狗，则它是哺乳动物。

如果上述说法正确，那么以下哪个说法一定正确？

(A) 如果一个动物不是哺乳动物，则它不是狗。

(B) 不是狗的动物都不是哺乳动物。

例题答案为选项 A 。我们可以先将原文转化为命题形式并绘制韦恩图。

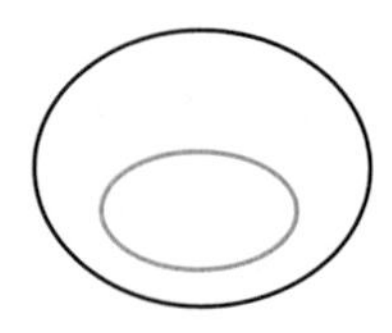

所有的狗都是哺乳动物。(图中橘色圈代表狗，黑色圈代表哺乳动物。)

选项 A 的意思是“黑色圈外的动物都不属于橘色圈之内”，这显然是正确的。选项 B 的意思是“橘色圈之外的动物都不在黑色圈之内”，这显然是不正确的。

确定结论类考题常见的问法有:

(1) Which of the following can be reliably concluded on the basis of the statements given?

(2) Among the following statements, which is it most reasonable to infer from the assertions by Mashika and Salim?

(3) Which of the following must be true in Greenspace County if the statements above are true?

这类考题的解法是非常简单的:

充分理解文段并且不加入任何自己的知识与想法，之后仔细检查选项是否被文段信息蕴含。

例题 1

From 1980 to 1989, total consumption of fish in the country of Jurania increased by 4.5 percent, and total consumption of poultry products there increased by 9.0 percent. During the same period, the population of Jurania increased by 6 percent, in part due to immigration to Jurania from other countries in the region.

If the statements above are true, which of the following must also be true on the basis of them?

(A) During the 1980s in Jurania, profits of wholesale distributors of poultry products increased at a greater rate than did profits of wholesale distributors of fish.

(B) For people who immigrated to Jurania during the 1980s, fish was less likely to be a major part of their diet than was poultry.

(C) In 1989 Juranians consumed twice as much poultry as fish.

(D) For a significant proportion of Jurania's population, both fish and poultry products were a regular part of their diet during the 1980s.

(E) Per capita consumption of fish in Jurania was lower in 1989 than in 1980.

推理：因为问题问的是 must be true，所以本题在考查确定结论。

选项分析

(A) 在 20 世纪 80 年代的 Jurania，家禽批发商获得的利润比鱼肉批发商获得的利润上涨得快。销量高不代表利润高，原文没有提到过利润。

(B) 对于 20 世纪 80 年代移民到 Jurania 的人来说，鱼比家禽更不可能成为他们的主要食材。移民之后，我们能看到的是食材消耗量的变化，但是并不代表原有居民的口味就不变，所以鱼的消耗量上涨得没有家禽的消耗量快不等于移民的人都喜欢吃家禽。

(C) 1989 年 Jurania 人吃的家禽量是吃的鱼量的两倍。1980 ~ 1989 年这段时间鱼和家禽消耗量的增长率差值是两倍，和总共消耗的数量是两个概念。

(D) 对于大部分的 Jurania 人来说，鱼和家禽都是他们在 20 世纪 80 年代的主要食材。该信息原文完全没有涉及。原文一直在说的是 Jurania 人口增长的情况，完全无法得出鱼和家禽是否是他们的主要食材。

（E）正确。1989 年人均消耗的鱼的数量比 1980 年的要低。因为鱼的消耗量增长了 4. 5%，而人口增加了 6%，所以平均每个人消耗的鱼的数量肯定是要下降的。（如果鱼的消耗量也增长 6%，则平均每个人消耗的鱼的数量可以不变。）

例题 2

In an attempt to genetically engineer a coffee plant that would produce beans containing no caffeine, scientists prevented the production of an enzyme necessary for the synthesis of caffeine. Beans harvested from plants in which production of the enzyme was shut down throughout the plant contained no caffeine. However, there were normal amounts of caffeine in beans harvested from plants in which production of the enzyme was shut down in beans but not in the rest of the plant.

If the information presented above is accurate, which of the following hypothesis is most strongly supported on the basis of it?

（A）Measurable amounts of caffeine are present in the leaves of a coffee plants in which production of the enzyme has been completely stopped.

（B）Caffeine production in coffee plants does not require the action of more than one enzyme.

（C）Coffee made from the beans of the plants in which enzyme production was shut down only in the beans contains both caffeine and small quantities of the enzyme.

（D）In coffee plants, either caffeine or the enzyme necessary for the production of caffeine moves into the beans from elsewhere in the plant.

(E) When the production of the enzyme is shut down in the beans but not in the rest of the coffee plant, the quantities of the enzyme produced in the rest of the plant are smaller than usual.

文段大意：科学家阻止了合成咖啡因所需的一种酶的产生。如果在整株植物中都阻止这种酶的产生，那么我们能收获不含咖啡因的豆子。但是，如果只是在豆子中阻止这种酶的产生，那么豆子中就会含有正常数量的咖啡因。

推理：因为问题问的是 be supported，所以本题在考查确定结论。

选项分析

(A) 在酶的产生已经完全停止的咖啡植物的叶片中存在着可测量数量的咖啡因。推理文段中没有提及咖啡植物叶片中是否含有咖啡因。

(B) 咖啡植物中咖啡因的产生不需要一种以上的酶的作用。推理文段只讲到了某一种酶是咖啡豆中合成咖啡因的必要条件，没有提及是否还需要别的酶参与。

(C) 用只有咖啡豆停止产生酶的植物的豆子制成的咖啡既含有咖啡因又含有少量的酶。我们只能确定这样的咖啡豆中必然会含有咖啡因，但推理文段中并未提及咖啡豆中是否会含有酶。

(D) 正确。在咖啡植物中，咖啡因或产生咖啡因所需的酶从植物的其他部分转移到咖啡豆中。显然本选项是文章蕴含的信息，即在咖啡豆中停止酶的合成的豆子里的咖啡因必然是从其他地方得来的，不可能凭空产生。

(E) 当咖啡豆中的酶停止产生，而咖啡植物的其他部分的酶没有停止产生时，植物的其他部分产生的酶的数量就会比平时少。推理文段只说了这样的咖啡豆含有的咖啡因的量和普通豆子是一样的，并没有谈到这种酶本身有多少的问题（我们也无法从推理文段的信息中确定酶和咖啡因的数量呈正比，也可能是无论产生多少咖啡因，都只需要等量的酶）。

例题3

Book publishing executive: I am becoming increasingly convinced that the interaction between book sales and literary quality is not a simple one. For example, I now believe that no number one best seller will win the major literary prizes. But I also believe that all major literary prize winners will have unusually high long-term sales.

If the beliefs that the book publishing executive claims to hold are true, which of the following must also be true on the basis of them?

(A) No books with unusually high long-term sales are number one best sellers.

(B) At least some number one best sellers are not books with unusually high long-term sales.

(C) All books with unusually high long-term sales are number one best sellers.

(D) At least some books with unusually high long-term sales are not number one best sellers.

(E) No number one best sellers have unusually high long-term sales.

推理：因为问题问的是 must be true，所以本题在考查确定结论。

从文段中可以提炼出两条信息：

（1）没有任何一本排名第一的畅销书能赢得目前主流的文学奖。

（2）所有主流文学奖得主的长期销量都会异常高。

其中信息（1）转换为命题形式为：

（3）所有赢得目前主流文学奖的书都不是畅销书。

因此，将信息（1）绘制成韦恩图可得（橘色圈代表赢得目前主流文学奖的书，黑色圈代表排名第一的畅销书）：

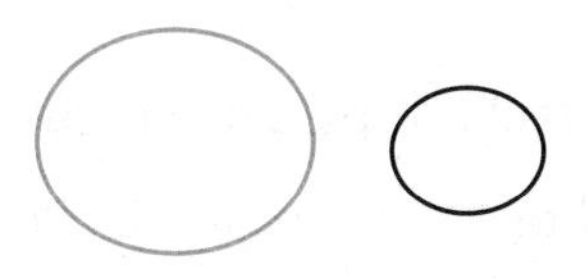

将信息（2）绘制成韦恩图可得（橘色圈代表赢得目前主流文学奖的书，黑色圈代表长期销量异常高的书）：

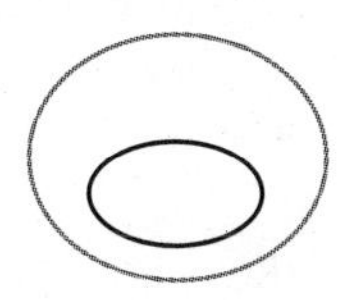

两条信息相结合可知，黑色圈和橘色圈的关系可能是黑色圈完全被橘色圈包含，也可能是黑色圈和橘色圈有交集，还有可能是黑色圈和橘色圈没有交集。

选项分析

（A）长期销量异常高的书都不是排名第一的畅销书。本选项讲的是橘色圈和黑色圈没有交集。显然，这一点无法从推理文段中确定。

（B）至少有些排名第一的畅销书不是长期销量异常高的书。本选项讲的是黑色圈不可能被橘色圈完全包含。显然无法确定。

（C）所有长期销量异常高的书都是排名第一的畅销书。本选项讲的是黑色圈包含橘色圈，这是不可能的。因为信息（1）告诉我们至少黑色圈不能包含橘色圈（而黑色圈在橘色圈里面）。

（D）正确。至少有些长期销售量异常高的书不是排名第一的畅销书。至少橘色圈不能被黑色圈包含。所以本选项是正确的。

（E）任何排名第一的畅销书的长期销量不会异常高。本选项讲的是黑色圈和橘色圈没有交集。错误同选项 A。

例题 4

That the application of new technology can increase the productivity of existing coal mines is demonstrated by the case of Tribnia's coal industry. Coal output per miner in Tribnia is double what it was five years ago, even though no new mines have opened.

Which of the following can be properly concluded from the statement about coal output per miner in the passage?

(A) If the number of miners working in Tribnian coal mines has remained constant in the past five years, Tribnia's total coal production has doubled in that period of time.

(B) Any individual Tribnian coal mine that achieved an increase in overall output in the past five years has also experienced an increase in output per miner.

(C) If any new coal mines had opened in Tribnia in the past five years, then the increase in output per miner would have been even greater than it actually was.

(D) If any individual Tribnian coal mine has not increased its output per miner in the past five years, then that mine's overall output has declined or remained constant.

(E) In Tribnia the cost of producing a given quantity of coal has declined over the past five years.

推理：因为问题问的是根据文段可以推出什么，所以本题在考查确定结论。

选项分析

(A) 正确。如果在 Tribnian 煤矿工作的工人数量没有变化的话，Tribnian 煤的总产量在同期应该上涨了两倍。推理文段指出，人均产量上涨两倍，且没有新矿，如果人数没有变化，那么由于“总产量 = 人均产量 × 人数”，所以总产量也应该上涨了两倍。

（B）任何一个在过去五年间总产量上涨的煤矿，人均产量也上涨了。总产量和人均产量以及人数的关系是无法从推理文段中得知的。

（C）如果过去五年间在 Tribnian 开采任何一个新的煤矿，那么人均开采量的增长会比它实际的要快。现有的煤矿人均开采量的增长不等于新煤矿人均开采量的增长。

（D）如果任何一个 Tribnian 煤矿的人均开采量在过去五年没有增长，那么其总量也一定会下降或者保持不变。总产量和人均产量以及人数的关系是无法从推理文段中得知的。

（E）在 Tribnian，生产一定数量的煤的成本在过去五年下降了。推理文段没有提到成本问题，产量高低与成本之间无必然关系。

例题 5

From 1973 to 1976, total United States consumption of cigarettes increased 3. 4 percent, and total sales of chewing tobacco rose 18. 0 percent. During the same period, total United States population increased 5. 0 percent.

If the statements above are true, which of the following conclusions can be properly drawn?

（A）United States manufacturers of tobacco products had higher profits in 1976 than in 1973.

（B）Per capita consumption of cigarettes in the United States was lower in 1976 than in 1973.

（C）The proportion of nonsmokers in the United States population dropped slightly between 1973 and 1976.

（D）United States manufacturers of tobacco products realize a lower profit on cigarettes than on chewing tobacco.

(E) A large percentage of United States smokers switched from cigarettes to chewing tobacco between 1973 and 1976.

推理：因为问题问的是根据文段可以推出什么，所以本题在考查确定结论。

选项分析

(A) 美国烟草制造商1976年的利润比1973年要高。推理文段中没有提到过关于利润的问题，所以本选项不是答案。

(B) 正确。1976年平均每人消耗的香烟量比1973年低。人口增加5%，而香烟消耗量只增加3.4%，所以本选项给出的结论正确。

(C) 1973—1976年，不抽烟的人的比例下降了。虽然chewing tobacco的消耗量显著上升，但是没有理由相信有更多人抽烟了（有可能是吸烟者每人抽的烟更多了）。

(D) 美国烟草制造商意识到香烟的利润低于嚼烟的利润。推理文段中没有提到过关于利润的问题，所以本选项不是答案。

(E) 1973—1976年，美国有很大比例的抽烟者从抽香烟转成了嚼烟。虽然chewing tobacco的消耗量迅速上升，但是由于香烟和嚼烟之间并不一定是此消彼长的关系（有可能既抽香烟又嚼烟），所以本选项不能保证从推理文段中推出。

2.2.2 确定前提

顾名思义，确定前提类考题会给出推理过程，但删除某个或某些前提，让我们选择填补前提的选项作为答案。这种问法的考题既可以考查我们对演绎论证的理解，又可以考查我们对假说论证的理解。例如：

小明每周三一定会穿衬衫。因此，小明今天会穿衬衫。

上述论证基于下列哪项假设？

(A) 今天是周三。

(B) 小明喜欢穿衬衫。

例题答案为选项A。显然，选项A补全了推理文段，使之成为"演绎论证"。又例如：

> 前提：家里的布局十分奇怪。
>
> 结论：肯定是有小偷来过了。
>
> 问题：上述论证基于下列哪项假设？
>
> （A）这种奇怪的布局可以被认为是一种人为破坏的杂乱。
>
> （B）再和谐的社会，也会存在小偷。

例题答案为选项A。在这道例题中，虽然我们补上的前提依然不能保证其结论是必然的，但选项A的成立让原论证成为更有效的“假说论证”。

> 回答基于假说论证的假设类考题应基于如下步骤：
>
> （1）判断前提和结论
>
> （2）识别推理类型
>
> （3）根据该类型对应的维度选出答案

我们将在“2.3　评估论证”中更详细地阐述这些步骤，此处仅以“因果推理”为例进行简单说明。

> 前提：我们观察到，凡是考GMAT的人都聪明。
>
> 结论：考GMAT导致人聪明。
>
> 问题：以上推理基于下列哪项假设？
>
> （A）不是聪明人才去考GMAT。
>
> （B）GMAT不是一个简单的考试。

根据结论和前提的特点，显然，该论证为因果推理。其共有三个评估维度：

（1）不存在纯粹巧合

（2）不存在因果倒置

（3）不存在他因导致结果

显然，该题目的答案为选项A，符合“因果倒置”这一维度。

常见的确定前提类考题的问法有：

（1）Which of the following is an assumption the researchers' reasoning requires?

（2）The personnel director's conclusion assumes which of the following?

（3）The conclusion above would be more reasonably drawn if which of the following were inserted into the argument as an additional premise?

例题 1

In an experiment, one group of volunteers were shown words associated with money, such as "salary", whereas another group was shown neutral words. Afterwards, individuals in both groups solved puzzles unrelated to money. Those who had been shown words associated with money were much less likely to request or offer help with the puzzles. The researchers concluded from this evidence that preoccupation with money makes people less cooperative.

Which of the following is an assumption the researchers' reasoning requires?

（A）At least some of the volunteers were preoccupied with money before being shown the words.

（B） Being from the neutral words did not cause the volunteers to become preoccupied with subjects other than money.

（C） Most of the volunteers who were shown neutral words requested or offered help with the puzzles.

（D） Most of the volunteers in both groups succeeded in solving the puzzles, either with or without help.

（E） The volunteers who were shown neutral words were, on average, less preoccupied with money while solving the puzzles than the other volunteers were.

前提：那些被展示过与金钱有关的词语的人，在解决难题时请求或提供帮助的可能性要小得多。

结论：对金钱的关注使人们不太愿意合作。

推理：前提中“金钱”和“合作”存在正相关的关系，且结论中两者为因果关系，因此，本题是因果推理。答案选项必然符合下面三个维度之一：

（1）不存在纯粹巧合

（2）不存在因果倒置

（3）不存在他因导致结果

选项分析

（A）至少有一些志愿者在被展示这些词之前就关注金钱。

（B）志愿者们在看到中性词的时候并没有对金钱以外的话题产生关注。

（C）大多数被展示了中性词的志愿者都请求或提供解谜题的帮助。其实我们并不关心被展示了中性词的人中大部分是否会提供帮助。这些人中无论是80%还是20%提供了帮助，只要关注金钱的人中提供帮助的人的比例比这些人低，那么我们就依然可以说明“对金钱的关注使人们不太愿意合作”。

(D) 两组中的大多数志愿者都成功地解开了谜题，不管有没有得到帮助。

(E) 正确。与其他志愿者相比，被展示了中性词的志愿者在解谜题时对金钱的关注度平均较低。本选项指出了金钱与合作具有明显的相关性，属于“不存在纯粹巧合”的维度。

例题 2

Even though most universities retain the royalties from faculty members' inventions, the faculty members retain the royalties from books and articles they write. Therefore, faculty members should retain the royalties from the educational computer software they develop.

The conclusion above would be more reasonably drawn if which of the following were inserted into the argument as an additional premise?

(A) Royalties from inventions are higher than royalties from educational software programs.

(B) Faculty members are more likely to produce educational software programs than inventions.

(C) Inventions bring more prestige to universities than do books and articles.

(D) In the experience of most universities, educational software programs are more marketable than are books and articles.

(E) In terms of the criteria used to award royalties, educational software programs are more nearly comparable to books and articles than to inventions.

前提：教职员工发明的版税归大学所有；教职员工撰写的书籍和文章的版税归自己所有。

结论：教职员工应该保留他们开发的教育软件的版税。

推理：本题让我们添加一个前提。观察文段的前提和结论可知，这道题讨论的焦点应在于“教育软件”是属于“发明”还是属于“书籍或文章”。因为结论要求教育软件的版税归教职员工自己所有，所以我们必须添加前提“教育软件属于书籍或文章”，以使原文成为“演绎论证”。

选项分析

（A）发明的版税高于教育软件程序的版税。

（B）与发明相比，教职员工更有可能制作教育软件程序。

（C）发明比书籍和文章更能给大学带来声誉。

（D）根据大多数大学的经验，教育软件程序比书籍和文章更有市场。

（E）正确。就授予版税的标准而言，教育软件程序更接近于书籍和文章，而不是发明。

例题3

In a certain wildlife park, park rangers are able to track the movements of many rhinoceroses because those animals wear radio collars. When, as often happens, a collar slips off, it is put back on. Putting a collar on a rhinoceros involves immobilizing the animal by shooting it with a tranquilizer dart. Female rhinoceroses that have been frequently recollared have significantly lower fertility rates than uncollared females. Probably, therefore, some substance in the tranquilizer inhibits fertility.

Which of the following is an assumption on which the argument depends?

（A）The dose of tranquilizer delivered by a tranquilizer dart is large enough to give the rangers putting collars on rhinoceroses a generous margin of safety.

(B) The fertility rate of uncollared female rhinoceroses in the park has been increasing in the past few decades.

(C) Any stress that female rhinoceroses may suffer as a result of being immobilized and handled has little or no negative effect on their fertility.

(D) The male rhinoceroses in the wildlife park do not lose their collars as often as the park's female rhinoceroses do.

(E) The tranquilizer used in immobilizing rhinoceroses is the same as the tranquilizer used in working with other large mammals.

前提：频繁戴项圈的母犀牛的生育率明显低于不戴项圈的母犀牛。

结论：麻醉剂影响了犀牛的生育。

推理：前提中“麻醉剂”和“生育”存在负相关的关系且结论中两者为因果关系，因此，本题是因果推理。答案选项必然符合下面三个维度之一：

(1) 不存在纯粹巧合

(2) 不存在因果倒置

(3) 不存在他因导致结果

选项分析

(A) 麻醉镖所释放的麻醉剂量足够大，以保证公园管理员在给犀牛安装项圈时足够安全。与三个维度无关。

(B) 公园里没有安装项圈的母犀牛的生育率在近几十年已经上涨了很多。本选项讨论的是没有安装项圈的母犀牛的生育率情况，不是关于推理文段中这些生育率低的犀牛的情况，可以排除。

(C) 正确。由于被麻醉而产生的紧张不会给犀牛的生育率带来不好的影响。本选项排除了一个他因，即排除了紧张导致生育率下降的可能。

(D) 野生动物园中公犀牛项圈掉落的频率要远小于母犀牛项圈掉落的频率。与公犀牛无关。

(E) 用来麻醉犀牛的麻醉剂和用来麻醉其他大型动物的麻醉剂相同。与麻醉剂是否相同无关。

2.2.3 现象解释

顾名思义，现象解释类题就是让我们解释一个现象或者一个观点。其解题原理是找到选项中最符合"该现象的结论（原因）"或者"该观点的论据"。我们一定要抓准要解释的现象或观点，时刻不偏离它们。例如：

原文：家里很杂乱。

下列哪个选项可以解释为什么家里很杂乱？

（A）可能有小偷来过了。

（B）定期的深度保洁可以让家里不那么杂乱。

例题答案为选项 A 。在这道例题中，我们相当于对"家里很杂乱"这一论据提出了一个可能的假说。

常见的现象解释类考题的问法有：

（1）Which of the following，if true，would most help to explain the surprising phenomenon described above?

（2）Which of the following most logically completes the passage?

（3）Which of the following，if true，provides the strongest justification for the farmers' reluctance?

例题 1

The number of applications for teaching positions in Newtown's public schools was 5. 7 percent lower in 1993 than in 1985 and 5. 9 percent lower in 1994 than in 1985. Despite a steadily growing student population and an increasing number of teacher resignations，however，Newtown does not face a teacher shortage in the late 1990's.

Which of the following, if true, would contribute most to an explanation of the apparent discrepancy above?

(A) Many of Newtown's public school students do not graduate from high school.

(B) New housing developments planned for Newtown are slated for occupancy in 1997 and are expected to increase the number of elementary school students in Newtown's public schools by 12 percent.

(C) The Newtown school board does not contemplate increasing the ratio of students to teachers in the 1990's.

(D) Teachers' colleges in and near Newtown produced fewer graduates in 1994 than in1993.

(E) In 1993 Newtown's public schools received 40 percent more applications for teaching positions than there were positions available.

文段大意：1993 年应聘教师的人数比 1985 年下降了 5.7%，1994 年应聘教师的人数比 1985 年下降了 5.9%，且学生数量和辞职教师的数量都在上升，但是，Newtown 居然没有碰到教师短缺的现象。

推理：本题要求我们解释为什么“Newtown 没有面临教师短缺的问题”。

选项分析

(A) 许多 Newtown 公立学校的学生都没有从高中毕业。如果学生没毕业，那教师应该更短缺，无法解释为什么不缺教师。

(B) Newtown 新的住房在 1997 年会被安排入住，并且有望给 Newtown 公立小学的学生数量带来 12% 的增长。如果学生数量还在上升，那么 Newtown 教师短缺的现象就更加严重了。

(C) Newtown 的学校管理层没有周密考虑 20 世纪 90 年代上涨的学生与教师的比例。无论学校的管理层是否考虑过教师短缺的问题，都不会影响教师是否真正短缺。

(D) Newtown 及其附近的师范学校 1994 年的毕业生比 1993 年少。本选项解释了为什么 1994 年应聘的教师比 1993 年少，但是无法解释为何不缺教师。

(E) 正确。1993 年 Newtown 公立学校收到的教师职位申请超过了现有职位数量的 40%。若 1993 年收到的教师职位申请比现有职位的数量多了很多，那么这证明 1993 年教师数量是供大于求的，自然 Newtown 在 20 世纪 90 年代末也就不一定缺乏教师。

例题 2

Wolves generally avoid human settlements. For this reason, domestic sheep, though essentially easy prey for wolves, are not usually attacked by them. In Hylantia prior to 1910, farmers nevertheless lost considerable numbers of sheep to wolves each year. Attributing this to the large number of wolves, in 1910 the government began offering rewards to hunters for killing wolves. From 1910 to 1915, large numbers of wolves were killed. Yet wolf attacks on sheep increased significantly.

Which of the following, if true, most helps to explain the increase in wolf attacks on sheep?

(A) Populations of deer and other wild animals that wolves typically prey on increased significantly in numbers from 1910 to 1915.

(B) Prior to 1910, there were no legal restrictions in Hylantia on the hunting of wolves.

(C) After 1910 hunters shot and wounded a substantial number of wolves, thereby greatly diminishing these wolves' ability to prey on wild animals.

(D) Domestic sheep are significantly less able than most wild animals to defend themselves against wolf attacks.

(E) The systematic hunting of wolves encouraged by the program drove many wolves in Hylantia to migrate to remote mountain areas uninhabited by humans.

文段大意：狼通常会避开人类的居住地。由于这个原因，家养的羊虽然很容易成为狼的猎物，但通常不会受到狼的攻击。1910 年之前，Hylantia 农民每年都因狼而失去大量羊。1910 年，政府认为狼的数量过多，开始悬赏鼓励猎人去猎杀狼。1910—1915 年，大量的狼被捕杀。然而，狼对羊的袭击却显著增加了。

推理：本题要求我们解释“为什么狼反而更多地攻击羊”。

选项分析

(A) 1910—1915 年，鹿和其他被狼所捕食的动物的数量显著上升了。本选项没有提到为何狼攻击羊，可以排除。

(B) 1910 年以前，Hylantia 对捕狼没有颁布限制法令。本选项没有提到狼的问题，可以排除。

(C) 正确。1910 年以后，猎手们射伤很多狼，因此狼捕食野生动物的能力大幅下降。本选项给出了“猎杀狼”之后可能带来的一个后果，即狼不再有能力捕食野生动物，转而更多地捕食羊了。

(D) 圈养的羊在抵抗狼的攻击方面要远远逊色于野生动物。本选项没有提到狼为何会攻击羊，可以排除。

(E) 系统地捕杀狼导致狼群大量迁徙到没有人类居住的山里。如果本选项成立，那么狼应该更少了且更无法攻击羊了。

例题 3

Which of the following most logically completes the argument?

The growing popularity of computer-based activities was widely predicted to result in a corresponding decline in television viewing. Recent studies have found that, in the United States, people who own computers watch, on average, significantly less television than people who do not own computers. In itself, however, this finding does very little to show that computer use tends to reduce television viewing time, since _______.

(A) many people who watch little or no television do not own a computer

(B) even though most computer owners in the United States watch significantly less television than the national average, some computer owners watch far more television than the national average

(C) computer owners in the United States predominantly belong to demographic groups that have long been known to spend less time watching television than the population as a whole does

(D) many computer owners in the United States have enough leisure time that spending significant amounts of time on the computer still leaves ample time for watching television

(E) many people use their computers primarily for tasks such as correspondence that can be done more rapidly on the computer, and doing so leaves more leisure time for watching television

文段大意：人们普遍预测，电脑使用的日益普及将导致看电视的时间相应减少。最近的研究发现，在美国，拥有电脑的人平均看电视的时间比没有电脑的人少得多。然而，这一发现本身并不能说明使用电脑会减少看电视的时间，因为_______。

推理：本题要求我们解释“为何研究发现不能表明使用电脑会减少看电视的时间”。

选项分析

（A）许多很少看电视或者不看电视的人都没有电脑。不看电视的人没有电脑，不能说明“他们有电脑后，是否会进一步减少看电视的时间”。

（B）虽然大部分有电脑的人看电视的时间要远低于国家的平均值，但是有些有电脑的人看电视的时间远远超过国家的平均值。某些国家的情况无法解释用电脑的人是否会减少看电视的时间。

（C）正确。美国有电脑的人大部分属于很少看电视的人群。本选项给出了“看电视时间短”的另外一个原因，即不是因为使用电脑，而是本身就不爱看电视。因此，不是因为使用电脑而看电视时间下降。

（D）美国许多有电脑的人有充足的业余时间，在电脑上花费大量时间，还依然有足够的时间来看电视。推理文段的前提部分已经讲过，用电脑的人看电视的时间短。也就是说，本选项只是讲了这部分人在什么时间看电视，无法解释为什么使用电脑会导致看电视的时间下降。

（E）许多人主要用电脑完成诸如通信之类的任务，这些任务可以在电脑上更快捷地完成，这样会留出更多空闲时间来看电视。本选项和 D 选项错误相同。

例题 4

Which of the following most logically completes the passage?

Using new detection techniques, researchers have found trace amounts of various medicinal substances in lakes and rivers. Taken in large quantities, these substances could have serious health effects, but they are present in quantities far too low to cause any physiological response in people who drink the water or bathe in it. Nevertheless, medical experts contend that eliminating these trace amounts from the water will have public health benefits, since ________.

(A) some of the medicinal substances found in lakes and rivers are harmless to humans even if taken in large quantities

(B) some of the medicinal substances found in lakes and rivers can counteract possible harmful effects of other such substances found there

(C) people who develop undesirable side effects when being treated with medicines that contain these substances generally have their treatment changed

(D) most medicinal substances that reach lakes or rivers rapidly break down into harmless substances

(E) disease-causing bacteria exposed to low concentrations of certain medicinal substances can become resistant to them

文段大意：利用新的检测技术，研究人员在湖泊和河流中发现了微量的各种药用物质。大量摄入这些物质可能会对健康造成严重影响，但它们的含量太低，在饮用或用这种水沐浴的人身上不会产生任何生理反应。然而，医学专家认为，去除水中的这些微量物质将对公众健康有益，因为________。

推理：本题要求我们解释“为何尽管这些物质含量低，但为了公众健康还是要去除”。

选项分析

(A) 就算大剂量摄入，水中的有些药物也不会对人体造成伤害。无论大量摄入是否会对人体造成伤害，本选项都和结论中的去除这些成分无关。

(B) 水中的有些药物可以中和其他此类药物可能产生的有害影响。本选项讲的是用什么方法可以去除那些药物，但不能解释为何要去除。

(C) 当使用含有这些物质的药物治疗时，那些出现不良副作用的人通常会改变治疗方案。本选项不能解释为何要去除这些成分。

(D) 大部分到达河流和湖泊的药物会迅速分解成无害物质。本选项提及了“药物”，但描述的是药物的原理。

(E) 正确。致病细菌长期暴露在某些低浓度的药物下会对它们产生抗药性。本选项讲“药物”的一个特殊特点，这个特点会导致无论水中的这些药物含量多少，去除这些药物，都会有利于公共卫生。

例题 5

Which of the following most logically completes the passage?

A certain tropical island received food donations in the form of powdered milk for distribution to its poorest residents, who were thought to be malnourished. Subsequently, the rate of liver cancers among those islanders increased sharply. The donated milk was probably to blame: recent laboratory research on rats has shown that rats briefly exposed to the substance aflatoxin tend to develop liver cancer when fed casein, a milk protein. This result is relevant because _______.

(A) in the tropics, peanuts, a staple of these island residents, support a mold growth that produces aflatoxin

(B) the liver is more sensitive to carcinogens, of which aflatoxin may be one, than most other bodily organs

(C) casein is not the only protein contained in milk

(D) powdered milk is the most appropriate form in which to send milk to a tropical destination

(E) the people who were given the donated milk had been screened for their ability to digest milk

文段大意：某热带岛屿收到了奶粉捐赠，分发给被认为营养不良的最贫穷的居民。随后，这些岛民的肝癌发病率急剧上升。捐赠的牛奶可能是罪魁祸首：最近对老鼠的实验室研究表明，短暂接触黄曲霉毒素的老鼠吃了酪蛋白（一种牛奶蛋白）容易患上肝癌。这个结果是相关的，因为________。

推理：本题选项只要能建立黄曲霉毒素和热带人之间的联系即可。

选项分析

（A）正确。在热带地区，花生——那里人的一种主食，（食用发霉的花生）会在身体里产生黄曲霉毒素。

（B）肝脏比其他器官更易受到像黄曲霉毒素这类致癌物质的影响。

（C）牛奶不仅仅含有酪蛋白一种蛋白质。

（D）奶粉是将牛奶送到热带地区最合适的方式。

（E）接受捐赠牛奶的人已经接受牛奶消化能力的检查。

练 习

请注意，所有基于假说论证的考题可以在学习“2. 3 评估论证”之后再进行练习。

1. AdadisCo needs to increase its annual production of sweatshirts. To do so, it must either hire new employees or else ask its current workforce to work more overtime. When working overtime, employees are paid one and one-half times their normal hourly wage. On the other hand, newly hired workers, although they would not be paid quite as much per hour as AdadisCo's experienced employees, would produce far less per hour than its current workers do.

Which of the following is most strongly supported by information given?

(A) Any new workers that AdadisCo hires would be trained by experienced current employees.

(B) AdadisCo cannot increase its production without experiencing at least a short-term increase in the average cost of labor per sweetshirt produced.

(C) If AdadisCo increases its annual production of sweatshirts, the quality of sweatshirts produced would decline.

(D) Not all of the employees in AdadisCo's current workforce would be willing to work more overtime if asked to do so.

(E) Unless AdadisCo asks its current workforce to work more overtime, its annual production will fall.

2. Last year, for the fourth year in a row, Sudeki's Bike Shop in Portland, Oregon, sold more TrackMaker bicycles than did any other single-location store in the state of Oregon. Sudeki's Bike Shop was also ranked third in the United States last year among single-location stores in terms of the number of TrackMaker bicycles sold, a rank it had never before attained.

If the statements above are true, which of the following CANNOT be true?

(A) Sudeki's Bike Shop sold fewer TrackMaker bicycles last year than the year before.

(B) More TrackMaker bicycles were sold in the state of Oregon than in any other state in the United States.

(C) Last year Sudeki's Bike Shop did not sell as many bicycles of the TrackMaker brand as it did of each of the other brands it sells.

(D) A bicycle store operating in more than one location in the state of Oregon sold more TrackMaker bicycles last year than did Sudeki's Bike Shop.

(E) The single-location store that was ranked second in the United States last year in terms of the number of TrackMaker bicycles sold was also located in the state of Oregon.

3. Many major corporations require their employees to wear conventional business attire, though they sometimes allow employees to wear casual attire on Fridays. Employees typically accomplish more work on such Fridays, and so their productivity is higher during those weeks than during weeks in which they must wear conventional attire every day. However, employees in corporations that allow casual attire every day are typically less productive than employees of corporations that require conventional attire every day.

The statements above, if true, can best be used as evidence supporting which of the following hypotheses?

(A) People's productivity at work is sometimes improved by variations in their work conditions.

(B) When people have more control over their work environments, they are more receptive to requests to change their work habits.

(C) Less work is generally performed on Fridays than is performed on other business days.

(D) People tend to work more productively when fewer restrictions are imposed on them.

(E) People are generally unable to change their work habits unless they are offered strong incentives to do so.

4. The number of vehicles built per year at certain automobile assembly plants has steadily increased over the past five years. In each of those years, the number of workers and worker productivity (number of automobiles produced per hour worked) have increased. The plant manager predicts that at least for the next two years, worker productivity will increase at the same rate as it has been increasing, and that over the next year the number of workers will not change.

If the statements above are true and the manager's predictions are accurate, which of the following is most strongly supported?

(A) Over the past five years, the increase in the number of vehicles built per year at the plant was greater than at most other automobile assembly plants.

(B) The number of vehicles built per year at the plant will begin to decrease after the next year.

(C) On average, workers at the plant will work longer hours next year than this year.

(D) The number of vehicles built per year at the plant will not increase at as fast a rate next year as it has over the past year.

(E) The average wage of the workers at the plant will decline next year.

5. In general, a business benefits when its customers feel they have strong ties to the employees they deal with. A customer is less likely to complain about a

service failure when the tie between customer and employee is strong. This is true regardless of whether the customer views the employee as responsible for the failure, presumably because the customer wishes to protect the employee from any negative consequences. But this has a downside. Businesses that do not get feedback in the form of customer complaints can be caught in a downward spiral of deteriorating quality. To avoid this outcome, businesses should ________.

Which of the following would most logically complete the passage above?

(A) discourage their employees from establishing strong ties with customers

(B) make refunds to any customer who complains even when the complaint is unwarranted

(C) provide a way for their customers to submit complains without identifying employees

(D) provide their employees with training in the effective handling of customer complaints

(E) present awards to employees who incur the least amount of customer dissatisfaction, as measured by surveys completed by customers

6. For a consumer product such as a television, careful pricing is a very important aspect of marketing, since pricing an item even slightly too high or too low can seriously reduce profits. By contrast, voter approval of government projects that benefit the public does not vary with small differences in estimates of their projected cost.

The statements above, if true, best support which of the following as a conclusion?

(A) Most people are well informed about the prices of consumer products.

(B) The demand for projects that benefit the public is more closely tied to standard measures of the condition of the national economy than is the demand for consumer products.

(C) Anyone wishing to increase or decrease voter support for projects that benefit the public should not focus on small cost differences.

(D) Many people place higher priority on funding projects that benefit the public than on buying consumer products for their households.

(E) The purchase of consumer products such as televisions can be postponed more easily than can expenditures for projects that benefit the public.

7. Apec's president said that payments made by Apec for health insurance constituted a smaller proportion of the employees' total compensation in 1999 than it had in 1994. The company's president went on to say that during this period, the terms on which Apec's employees received medical insurance benefits did not change.

If the statements attributed to Apec's president are true, which of the following must also be true about Apec in the period between 1994 and 1999?

(A) Nonsalary benefits other than health insurance that Apec provides for its employees also constituted a decreasing proportion of the employees' total compensation.

(B) There was no increase in the costs, to the insurance company or companies, of covering the health-care needs of Apec's employees, or if there was, it was not passed on to Apec.

(C) The productivity of Apec's employees, computed as the per-employee share of the company's total annual output, increased.

(D) Either the cost to Apec of providing health insurance for its employees decreased or the total compensation received by Apec's employees increased, or both.

(E) Apec's net profits went up each year throughout this period

8. In Nurica, investors who believe they are the victims of malpractice by their financial advisors can file a claim with a government agency. The agency arbitrates the dispute unless a settlement is reached before the hearing date. Of those claims that reached the arbitration stage, a smaller proportion resulted in restitution for the investor in 1995 than in 1994. Nonetheless, a larger proportion of all the claims filed resulted in restitution in 1995 than in 1994.

Which of the following is most strongly supported by the information given?

(A) The proportion of claims filed that were unfounded was higher in 1995 than in 1994.

(B) The average amount awarded to investors whose claims were arbitrated by the agency was lower in 1995 than in 1994.

(C) The average amount of restitution for settlement was higher in 1995 than in 1994.

(D) There were fewer arbitrators available to hear claims in 1995 than in 1994.

(E) A larger proportion of claims settled prior to arbitration resulted in restitution in 1995 than in 1994.

9. A preliminary announcement from the government of the Republic of Bilong indicated that the country's year-to-date trade deficit dropped again last month. In fact, according to a government spokesperson, last month was the third month in a row that Bilong's exports were substantially greater in value than its imports. The government has not yet released any actual figures, but clearly the trend reported indicates that government efforts to promote Bilong's export trade are finally bearing fruit.

Which of the following is an assumption on which the argument depends?

(A) There has not been a sharp contraction in the total value of goods imported into Bilong in recent months.

(B) There has been no lessening in recent months in the government's efforts to promote exports.

(C) The government's efforts to promote export trade did not lead Bilong's exporters to reduce prices in order to achieve a higher volume of exports.

(D) Bilong's government is able to discern broad trends in foreign trade long before actual figures are available.

(E) There has been no significant shift in Bilong's exports from one category of exported goods to another in recent months.

10. Research shows that over half of all morning political talk shows on the major television networks do not include any women broadcasters, and in terms of total guest appearances on these shows, women represent just 14% of the guests. Why does this research matter? These shows are a critical forum for our nation's political debates. They have an agenda-setting effect, in that they tend to influence priorities of citizens across the country. The relative lack of women on these shows results in women's issues being a lower political priority for most citizens.

The argument above relies on which of the following assumptions?

(A) Most of the nation's citizens watch morning news talk shows.

(B) Morning news talk shows exhibit greater gender disparities than other news programs.

(C) Women's issues will not be a political priority for people who do not see them discussed on news talk shows.

(D) News talk shows are the primary mechanism that shapes the nation's political debates.

(E) Hosts and guests of news talk shows are more likely to raise issues that are pertinent specifically to their own gender.

第二组

1. Allegretto Winery is a small winery whose wines are sought after by collectors. At the end of this year's grape-growing season, there was a long hot dry spell, which kept the grapes small. Since Allegretto uses only grapes grown in its own vineyard, it expects to produce significantly less wine than usual. Nevertheless, Allegretto expects that its revenue from this year's wine will equal its revenue from more normal years.

Which of the following, if true, does most to justify the winery's expectation?

(A) Small grapes do not require significantly less labor to harvest than normal-sized grapes do.

(B) Long hot dry spells at the beginning of the grape-growing season are rare, but they can have a devastating effect on a vineyard's yield.

(C) Many wineries make wine from grapes grown in vineyards they do not own, sometimes even from grapes grown in other regions.

(D) When hot dry spells are followed by heavy rains, the rains frequently destroy grape crops.

(E) Grapes that have matured in hot dry weather make significantly better wine than ordinary grapes do.

2. Which of the following most logically completes the argument below?

The Republic of Naboa is replacing its $2 bills with $2 coins. On average, each coin costs more to produce than does each bill. Nevertheless, producing coins instead of bills will enable the government of Naboa to save substantially on its production costs for $2 currency in future years, since _______.

(A) the coins will be introduced gradually into circulation as worn $2 bills are taken out of circulation

(B) all currency in Naboa of denominations higher than $2 will continue to be paper currency

(C) consumers are no less likely to use a $2 coin than they are to use a $2 bill

(D) the total amount of the metal used per $2 coin is itself valued at much less than $2

(E) the coins will remain in circulation for an average of twenty years, whereas $2 bills generally wear out after only two years

3. Over the past several years coffee consumption in the United States has declined while the total combined revenues of American coffee companies have increased substantially. Although all coffee companies have increased prices on all brands of coffee they sell, the increase in revenues is much higher than can be accounted for by the increase in prices.

Which of the following, if true, most contributes to an explanation of the increase in revenues not accounted for by the increase in prices?

(A) The decline in coffee consumption has forced many smaller coffee companies to sell their operations to companies that are market leaders.

(B) Many former coffee drinkers have switched to drinking herbal teas and caffeine-free sodas.

(C) Many coffee drinkers have switched from bargain brands of coffee to more expensive premium brands.

(D) Coffee-exporting countries have placed larger export tariffs on coffee to compensate for declining sales.

(E) New brewing technologies allow coffee drinkers to use less ground coffee per cup of brewed coffee.

4. Which of the following most logically completes the argument?

The Bluewater Hotel usually earns half its yearly revenue during the year's first quarter, when typically all its rooms are occupied. At other times of the year it has many vacancies. This year the Bluewater was closed throughout the first quarter while damage caused by a severe storm at the end of last year was repaired. Nevertheless, the Bluewater is unlikely to see this year's revenue drop to half of a normal year's revenue, since ________.

(A) the closure this year will not affect its revenue next year, provided the Bluewater is open during next year's first quarter

(B) few of the other hotels in the region were forced to close for repairs after the storm

(C) the cost of the repairs was less than half of a normal year's revenue

(D) it is extremely unlikely that any storm this year will be so severe as to cause damage comparable to the damage caused by the storm last year

(E) many of the customers who would ordinarily have stayed at the Bluewater during the first quarter will probably reschedule their stay for later in the year

5. The country of Karlandia imposes a tax of three cents per gallon on sales of gasoline. The revenues from this tax are used to maintain Westland's roads. The number of miles driven per year in Karlandia has doubled since the tax went into effect fifteen years ago, but although taxes are collected as diligently as ever, gasoline tax revenues have not doubled.

Which of the following, if true, best reconciles the apparent discrepancy?

(A) Many miles of new roads, all of which will require regular maintenance, are under construction in Westland.

(B) Overall, vehicles in Karlandia consume less fuel per mile driven than they did 15 years ago.

(C) Gasoline distributors in Karlandia import more gasoline than they did fifteen years ago.

(D) The number of vehicles in Karlandia has increased dramatically in the last fifteen years.

(E) The price of gasoline has not doubled in the last fifteen years.

6. Sales of Paradoze, a nonprescription sleeping pill, declined throughout the six months in which a certain television advertisement was used to promote the product. Yet the makers of Paradoze consider the television advertisement to have been very effective.

Which of the following, if true, most helps to resolve the apparent discrepancy?

(A) The makers of Paradoze had previously used television advertising to promote nonprescription products.

(B) On average, only one out of four advertisements for a nonprescription drug is judged effective by the drug's manufacturer.

(C) During the six months in question, Paradoze's share of a shrinking market for nonprescription sleeping pills increased.

(D) Paradoze was the only nonprescription sleeping pill being promoted through television advertisements during the six months in question.

(E) Paradoze's share of the market for nonprescription sleeping pills has always been small in comparison with that of its largest competitor.

7. In Kravonia, the average salary for jobs requiring a college degree has always been higher than the average salary for jobs that do not require a degree. Current enrollments in Kravonia's colleges indicate that over the next four years the percentage of the Kravonian workforce with college degrees will increase dramatically. Therefore, the average salary for all workers in Kravonia is likely to increase over the next four years.

Which of the following is an assumption on which the argument depends?

(A) Kravonians with more than one college degree earn more, on average, than do Kravonians with only one college degree.

(B) The percentage of Kravonians who attend college in order to earn higher salaries is higher now than it was several years ago.

(C) The higher average salary for jobs requiring a college degree is not due largely to a scarcity among the Kravonian workforce of people with a college degree.

(D) The average salary in Kravonia for jobs that do not require a college degree will not increase over the next four years.

(E) Few members of the Kravonian workforce earned their degrees in other countries.

8. Until now, only injectable vaccines against influenza have been available. Parents are reluctant to subject children to the pain of injections, but adults, who are at risk of serious complications from influenza, are commonly vaccinated. A new influenza vaccine, administered painlessly in a nasal spray, is effective for children. However, since children seldom develop serious complications from influenza, no significant public health benefit would result from widespread vaccination of children using the nasal spray.

Which of the following is an assumption on which the argument depends?

(A) Any person who has received the injectable vaccine can safely receive the nasal-spray vaccine as well.

(B) The new vaccine uses the same mechanism to ward off influenza as injectable vaccines do.

(C) The injectable vaccine is affordable for all adults.

(D) Adults do not contract influenza primarily from children who have influenza.

(E) The nasal spray vaccine is not effective when administered to adults.

9. Newspaper editorial: A surprising study by the World Health Organization found that chronic exposure to excessive noise may cause or exacerbate serious health problems such as stroke and heart disease. This study will produce health benefits for our city, for it will inevitably lead to pressure on our local politicians, who will respond by passing stricter noise ordinances.

The editorial's reasoning relies on which of the following assumptions?

(A) The most effective way to reduce the public's exposure to excessive noise is through stricter noise ordinances.

(B) Stricter noise ordinances will lessen the amount of excessive noise people would otherwise be exposed to.

(C) Rates of heart disease and stroke tend to be highest in areas where excessive noise is great.

(D) Local politicians who help pass stricter noise ordinances will do so primarily to improve the health of their constituents.

(E) The newspaper's city currently has less strict noise ordinances than do many other cities of comparable size.

10. For similar cars and drivers, automobile insurance for collision damage has always cost more in Greatport than in Fairmont. Police studies, however, show that cars owned by Greatport residents are, on average, slightly less likely to be involved in a collision than cars in Fairmont. Clearly, therefore, insurance companies are making a greater profit on collision-damage insurance in Greatport than in Fairmont.

Which of the following is an assumption on which the argument depends?

(A) Repairing typical collision damage does not cost more in Greatport than in Fairmont.

(B) There are no more motorists in Greatport than in Fairmont.

(C) Greatport residents who have been in a collision are more likely to report it to their insurance company than Fairmont residents are.

(D) Fairmont and Greatport are the cities with the highest collision-damage insurance rates.

(E) The insurance companies were already aware of the difference in the likelihood of collisions before the publication of the police reports.

答案及解析

第一组

1. AdadisCo needs to increase its annual production of sweatshirts. To do so, it must either hire new employees or else ask its current workforce to work more overtime. When working overtime, employees are paid one and one-half times their normal hourly wage. On the other hand, newly hired workers, although they would not be paid quite as much per hour as AdadisCo's experienced employees, would produce far less per hour than its current workers do.

Which of the following is most strongly supported by information given?

(A) Any new workers that AdadisCo hires would be trained by experienced current employees.

(B) AdadisCo cannot increase its production without experiencing at least a short-term increase in the average cost of labor per sweetshirt produced.

(C) If AdadisCo increases its annual production of sweatshirts, the quality of sweatshirts produced would decline.

(D) Not all of the employees in AdadisCo's current workforce would be willing to work more overtime if asked to do so.

(E) Unless AdadisCo asks its current workforce to work more overtime, its annual production will fall.

文段大意：AdadisCo 公司需要增加其运动衫的年产量。要做到这一点，它必须雇用新的员工，否则就要求现有的员工多加班。加班时，员工的工资是其正常小时工资的 1.5 倍。另一方面，新招聘的工人，尽管他们每小时的工资不会像 AdadisCo 公司有经验的员工那么多，但每小时的产量却远远低于现有工人的产量。

推理：问题问的是哪个选项可以被文段支持，所以本题在考查构建论证。

选项分析：

(A) AdadisCo 公司雇用的任何新工人都会接受有经验的现有员工的培训。文段没有提及新员工是否会接受老员工的培训。

(B) 正确。AdadisCo 公司在增加产量的同时，至少要经历生产每件运动衫的平均劳动成本在短期内增加。无论走哪条路，都会带来成本的增加。

(C) 如果 AdadisCo 公司增加运动衫的年产量，那么生产的运动衫的质量就会下降。文段完全没提质量的事情。

(D) 如果要求 AdadisCo 公司的员工加班，并不是所有的员工都会愿意加班。文段没有提加班意愿的事情。

(E) 除非 AdadisCo 公司要求其目前的员工多加班，否则其年产量会下降。和原文相悖，除了让现有员工加班外，还可以选择雇用新员工。

2. Last year, for the fourth year in a row, Sudeki's Bike Shop in Portland, Oregon, sold more TrackMaker bicycles than did any other single-location store in the state of Oregon. Sudeki's Bike Shop was also ranked third in the United States last year among single-location stores in terms of the number of TrackMaker bicycles sold, a rank it had never before attained.

If the statements above are true, which of the following CANNOT be true?

(A) Sudeki's Bike Shop sold fewer TrackMaker bicycles last year than the year before.

(B) More TrackMaker bicycles were sold in the state of Oregon than in any other state in the United States.

(C) Last year Sudeki's Bike Shop did not sell as many bicycles of the TrackMaker brand as it did of each of the other brands it sells.

(D) A bicycle store operating in more than one location in the state of Oregon sold more TrackMaker bicycles last year than did Sudeki's Bike Shop.

(E) The single-location store that was ranked second in the United States last year in terms of the number of TrackMaker bicycles sold was also located in the state of Oregon.

文段大意：去年，俄勒冈州波特兰市的 Sudeki 自行车店连续第四年卖出的 TrackMaker 自行车数量超过了俄勒冈州其他任何一家单一门店。去年，Sudeki 自行车店的 TrackMaker 自行车的销量也在美国单一门店中排名第三，这是它以前从未获得过的排名。

推理：问题问的是“哪个选项不可能是真的”，所以本题在反向考查构建论证。需要找一个成立可能性为 0 的选项。

选项分析：

(A) Sudeki 自行车店去年售出的 TrackMaker 自行车比前年少。原文没有比较去年和前年的销量。

(B) 俄勒冈州售出的 TrackMaker 自行车比美国其他任何州都多。原文没有以“州”为单位做任何比较。

(C) 去年 Sudeki 自行车店售出的 TrackMaker 自行车没有它所售出的其他品牌的自行车多。原文没有提及“其他品牌的自行车”。

(D) 俄勒冈州一家在多个地点经营的自行车店去年售出的 TrackMaker 自行车比 Sudeki 自行车店售出的多。原文没有提及“在多个地点经营的自行车店”。

(E) 正确。去年 TrackMaker 自行车销量在美国排名第二的单一门店也位于俄勒冈州。推理文段中讲到，俄勒冈州售出 TrackMaker 自行车最多的商店就是 Sudeki 自行车店，而 Sudeki 自行车店是全美第三，因此，不可能还有俄勒冈州的店能比 Sudeki 自行车店售出的 TrackMaker 自行车多，自然全美第二的店不可能在俄勒冈州。

3. Many major corporations require their employees to wear conventional business attire, though they sometimes allow employees to wear casual attire on Fridays. Employees typically accomplish more work on such Fridays, and so their productivity is higher during those weeks than during weeks in which they must

wear conventional attire every day. However, employees in corporations that allow casual attire every day are typically less productive than employees of corporations that require conventional attire every day.

The statements above, if true, can best be used as evidence supporting which of the following hypotheses?

(A) People's productivity at work is sometimes improved by variations in their work conditions.

(B) When people have more control over their work environments, they are more receptive to requests to change their work habits.

(C) Less work is generally performed on Fridays than is performed on other business days.

(D) People tend to work more productively when fewer restrictions are imposed on them.

(E) People are generally unable to change their work habits unless they are offered strong incentives to do so.

文段大意：许多大公司要求他们的员工穿传统职业装，尽管他们有时允许员工在星期五穿休闲装。员工通常在这样的星期五完成更多的工作，因此他们在这几周的生产力要比每天必须穿传统职业装的那几周高。然而，允许每天穿休闲装的公司的员工通常比要求每天穿传统职业装的公司的员工生产率低。

推理：问题问的是“文段可以作为证据，支持下面哪个假设”，所以本题在考查构建论证。

选项分析：

(A) 正确。人们在工作中的生产力有时会因工作条件的变化而提高。

(B) 当人们对他们的工作环境有更多控制时，他们更容易接受改变他们工作习惯的要求。原文没有提及“控制环境”和“接受要求”之间的关系。

(C) 星期五的工作一般比其他工作日的工作要少。原文没有提及工作量的事情。

（D）当人们受到的限制较少时，他们的工作往往更有成效。原文表示“每天穿休闲装，效率反而比每天穿传统职业装的时候低”，所以此选项无法根据文段得出。

（E）人们通常无法改变他们的工作习惯，除非提供强有力的激励。原文没有提如何才能改变工作习惯。

4. The number of vehicles built per year at certain automobile assembly plants has steadily increased over the past five years. In each of those years, the number of workers and worker productivity (number of automobiles produced per hour worked) have increased. The plant manager predicts that at least for the next two years, worker productivity will increase at the same rate as it has been increasing, and that over the next year the number of workers will not change.

If the statements above are true and the manager's predictions are accurate, which of the following is most strongly supported?

(A) Over the past five years, the increase in the number of vehicles built per year at the plant was greater than at most other automobile assembly plants.

(B) The number of vehicles built per year at the plant will begin to decrease after the next year.

(C) On average, workers at the plant will work longer hours next year than this year.

(D) The number of vehicles built per year at the plant will not increase at as fast a rate next year as it has over the past year.

(E) The average wage of the workers at the plant will decline next year.

文段大意：在过去的五年里，某些汽车装配厂每年制造的汽车数量一直在稳步增长。在这些年里，工人的数量和工人的生产力（每小时工作生产的汽车数量）都在增长。工厂经理预测，至少在未来两年内，工人的生产力将以同样的速度增长，而在未来一年内，工人的数量将不会改变。

推理：问题问的是哪个选项可以被文段内容支持，所以本题在考查构建论证。

选项分析：

(A) 在过去的五年里，该厂每年制造的汽车数量的增长幅度大于其他大多数汽车装配厂。文段没有提及其他汽车装配厂。

(B) 该厂每年制造的汽车数量将在明年以后开始减少。与原文信息不符。

(C) 平均来说，该厂的工人明年的工作时间将比今年长。文段没有提及工人工作时长的情况。

(D) 正确。明年该厂一年制造的汽车数量不会像过去一年那样快速增长。过去工人数量和工人生产力都在增长，而未来一年工人数量不变，所以未来一年工人生产力的增长肯定不如过去。

(E) 明年该厂工人的平均工资将下降。文段没有提及工资的情况。

5. In general, a business benefits when its customers feel they have strong ties to the employees they deal with. A customer is less likely to complain about a service failure when the tie between customer and employee is strong. This is true regardless of whether the customer views the employee as responsible for the failure, presumably because the customer wishes to protect the employee from any negative consequences. But this has a downside. Businesses that do not get feedback in the form of customer complaints can be caught in a downward spiral of deteriorating quality. To avoid this outcome, businesses should ________.

Which of the following would most logically complete the passage above?

(A) discourage their employees from establishing strong ties with customers

(B) make refunds to any customer who complains even when the complaint is unwarranted

(C) provide a way for their customers to submit complains without identifying employees

(D) provide their employees with training in the effective handling of customer complaints

(E) present awards to employees who incur the least amount of customer dissatisfaction, as measured by surveys completed by customers

文段大意：一般来说，当客户觉得他们和与他们打交道的员工有很强的联系时，企业就会受益。当客户和员工之间的联系紧密时，客户就不太可能抱怨服务失败。无论客户是否认为员工应对失败负责，这都是正确的，大概是因为客户希望保护员工免受任何负面后果的影响。但这也有缺点。没有受到客户投诉反馈的企业可能会陷入质量恶化的恶性循环。为了避免这种结果，企业应该________。

推理：问题问的是哪个选项使得原文逻辑成立，所以本题在考查构建论证。

选项分析：

(A) 不鼓励他们的员工与客户建立紧密的联系。客户与员工建立紧密联系是有好处的，所以不能不鼓励。

(B) 对任何投诉的客户进行退款，即使投诉是没有理由的。与退不退款无关。

(C) 正确。为他们的客户提供一种方法，使他们可以在不确定员工身份的情况下提交投诉。此方法结合了客户与员工建立联系的优缺点。

(D) 为其员工提供有效处理客户投诉的培训。与员工要如何处理客户投诉无关。

(E) 根据客户完成的调查，给那些引起客户不满程度最低的员工颁发奖励。此举动无法减少客户与员工建立联系的弊端。

6. For a consumer product such as a television, careful pricing is a very important aspect of marketing, since pricing an item even slightly too high or too low can seriously reduce profits. By contrast, voter approval of government projects that benefit the public does not vary with small differences in estimates of their projected cost.

The statements above, if true, best support which of the following as a conclusion?

(A) Most people are well informed about the prices of consumer products.

(B) The demand for projects that benefit the public is more closely tied to standard measures of the condition of the national economy than is the demand for consumer products.

(C) Anyone wishing to increase or decrease voter support for projects that benefit the public should not focus on small cost differences.

(D) Many people place higher priority on funding projects that benefit the public than on buying consumer products for their households.

(E) The purchase of consumer products such as televisions can be postponed more easily than can expenditures for projects that benefit the public.

文段大意：对于像电视机这样的消费品来说，谨慎定价是营销的一个非常重要的方面，因为一件商品的定价即使稍微过高或过低都会严重降低利润。相比之下，选民对有益于公众的政府项目的认可度并不因其预估成本的微小差异而变化。

推理：问题问的是文段最支持哪个选项，所以本题在考查构建论证。

选项分析：

(A) 大多数人对消费产品的价格都很了解。文段没有提及。

(B) 与对消费品的需求相比，对有益于公众的项目的需求与国民经济状况的衡量标准有更紧密的联系。原文没有提及需求的事情。

(C) 正确。任何希望增加或减少选民对有益于公众的项目的支持的人都不应该把注意力放在小的成本差异上。与原文一致。

(D) 与为他们的家庭购买消费品相比，许多人更重视资助有益于公众的项目。原文没有将两者的重要性做比较。

(E) 购买电视机等消费品比资助有益于公众的项目更容易推迟。原文没有提及。

7. Apec's president said that payments made by Apec for health insurance constituted a smaller proportion of the employees' total compensation in 1999 than it had in 1994. The company's president went on to say that during this

period, the terms on which Apec's employees received medical insurance benefits did not change.

If the statements attributed to Apec's president are true, which of the following must also be true about Apec in the period between 1994 and 1999?

(A) Nonsalary benefits other than health insurance that Apec provides for its employees also constituted a decreasing proportion of the employees' total compensation.

(B) There was no increase in the costs, to the insurance company or companies, of covering the health-care needs of Apec's employees, or if there was, it was not passed on to Apec.

(C) The productivity of Apec's employees, computed as the per-employee share of the company's total annual output, increased.

(D) Either the cost to Apec of providing health insurance for its employees decreased or the total compensation received by Apec's employees increased, or both.

(E) Apec's net profits went up each year throughout this period.

文段大意：Apec 公司总裁说，1999 年 Apec 公司支付的健康保险费用在员工总报酬中所占的比例比 1994 年要小。该公司总裁接着说，在此期间，Apec 公司员工获得医疗保险福利的条件并没有改变。

推理：问题问的是以下哪项一定是真的，所以本题在考查构建论证。

选项分析：

(A) Apec 公司为其员工提供的医疗保险以外的非工资性福利在员工的总报酬中所占的比例也在下降。

(B) 保险公司为 Apec 员工提供医疗服务的成本没有增加，或者即使增加了，也没有转嫁给 Apec 公司。

(C) Apec 员工的生产力（按每个员工在公司年度总产出中的份额计算）有所增加。

(D) 正确。要么 Apec 公司为其员工提供的健康保险费用减少，要么 Apec 公司的员工获得的总报酬增加，或者两者都有。

(E) 在这一时期，Apec 公司的净利润每年都在上升。

8. In Nurica, investors who believe they are the victims of malpractice by their financial advisors can file a claim with a government agency. The agency arbitrates the dispute unless a settlement is reached before the hearing date. Of those claims that reached the arbitration stage, a smaller proportion resulted in restitution for the investor in 1995 than in 1994. Nonetheless, a larger proportion of all the claims filed resulted in restitution in 1995 than in 1994.

Which of the following is most strongly supported by the information given?

(A) The proportion of claims filed that were unfounded was higher in 1995 than in 1994.

(B) The average amount awarded to investors whose claims were arbitrated by the agency was lower in 1995 than in 1994.

(C) The average amount of restitution for settlement was higher in 1995 than in 1994.

(D) There were fewer arbitrators available to hear claims in 1995 than in 1994.

(E) A larger proportion of claims settled prior to arbitration resulted in restitution in 1995 than in 1994.

文段大意：在努里卡，投资者如果认为自己是金融顾问不当行为的受害者，可以向政府机构提出索赔。除非在听证会日期前达成和解，否则该机构将对争议进行仲裁。在进入仲裁阶段的索赔中，1995 年投资者获得赔偿的比例比 1994 年小。尽管如此，1995 年提出的所有索赔中，获得赔偿的比例比 1994 年要大。

提出索赔 → 达成和解 → 赔偿

提出索赔 → 仲裁 → 赔偿

仲裁之后的赔偿只是所有赔偿中的一种，还有一种方式是在达成和解后产生的赔偿。既然1995年和1994年相比，仲裁之后的赔偿比例低，但总赔偿的比例高，那可能是因为达成和解后产生的赔偿比例高。

推理：问题问的是以下哪项可以被文段支持，所以本题在考查构建论证。

选项分析：

（A）1995年提出的索赔要求，没有根据地高于1994年。
（B）1995年经该机构仲裁的投资者获得的平均赔偿额低于1994年。
（C）1995年用于和解的平均赔偿金额高于1994年。
（D）1995年可审理索赔的仲裁员比1994年少。
（E）正确。1995年在仲裁前获得赔偿的比例比1994年大。

9. A preliminary announcement from the government of the Republic of Bilong indicated that the country's year-to-date trade deficit dropped again last month. In fact, according to a government spokesperson, last month was the third month in a row that Bilong's exports were substantially greater in value than its imports. The government has not yet released any actual figures, but clearly the trend reported indicates that government efforts to promote Bilong's export trade are finally bearing fruit.

Which of the following is an assumption on which the argument depends?

(A) There has not been a sharp contraction in the total value of goods imported into Bilong in recent months.
(B) There has been no lessening in recent months in the government's efforts to promote exports.
(C) The government's efforts to promote export trade did not lead Bilong's exporters to reduce prices in order to achieve a higher volume of exports.
(D) Bilong's government is able to discern broad trends in foreign trade long before actual figures are available.

(E) There has been no significant shift in Bilong's exports from one category of exported goods to another in recent months.

类别：普通预测推理

前提：上个月是连续第三个月，比隆的出口值大大高于进口值。

结论：政府对促进比隆出口贸易所做出的努力终于取得了成果。

推理：结论“对促进出口贸易所做出的努力取得了成果（出口提高了）”的充分条件应为：出口值大于进口值且进口本身并未减少。因此，答案选项必须指出“进口本身并未减少”。

选项分析：

(A) 正确。近几个月来，比隆进口的货物总值没有急剧收缩。

(B) 最近几个月，政府在促进出口方面所做出的努力没有减少。此选项相当于重复了一遍结论。

(C) 政府对促进出口贸易所做出的努力没有导致比隆的出口商为了实现更多的出口量而降低价格。此选项讨论的是出口商提高出口量的手段，与三个维度无关。

(D) 早在实际数据公布之前，比隆政府就能够觉察出对外贸易的大趋势。与政府能否察觉出趋势无关。

(E) 最近几个月，比隆的出口没有从一类出口商品显著转向另一类出口商品。与出口商品的类别无关。

10. Research shows that over half of all morning political talk shows on the major television networks do not include any women broadcasters, and in terms of total guest appearances on these shows, women represent just 14% of the guests. Why does this research matter? These shows are a critical forum for our nation's political debates. They have an agenda-setting effect, in that they tend to influence priorities of citizens across the country. The relative lack of

women on these shows results in women's issues being a lower political priority for most citizens.

The argument above relies on which of the following assumptions?

(A) Most of the nation's citizens watch morning news talk shows.

(B) Morning news talk shows exhibit greater gender disparities than other news programs.

(C) Women's issues will not be a political priority for people who do not see them discussed on news talk shows.

(D) News talk shows are the primary mechanism that shapes the nation's political debates.

(E) Hosts and guests of news talk shows are more likely to raise issues that are pertinent specifically to their own gender.

类别：普通预测推理

前提：主要电视网络上所有早间政治谈话节目中，超过一半的节目没有任何女性播音员。并且，就这些节目的总出场嘉宾而言，女性仅占14%。

结论：对大多数公民来说，女性事务的政治优先地位较低。

推理：结论“女性事务的政治优先地位较低”的充分条件应为：女性事务在节目中很少出现。（原文明确给出了政治谈话节目和政治优先地位是高度相关的。）因此，答案选项必须指出“女性嘉宾的占比和女性事务在政治谈话节目中的占比是有关甚至相等的”。

选项分析：

(A) 大多数公民都看早间新闻谈话节目。和公民看不看节目无关。

(B) 早间新闻谈话节目比其他新闻节目表现出更大的性别差异。与和别的节目相比较无关。

(C) 对那些没有在新闻谈话节目中看到女性事务被讨论的人来说，女性事务不会享有政治优先权。结论讨论的是“对于大多数公民来说”，但我们并不知道没有在新闻谈话节目中看到女性事务被讨论的人占多少，所以此选项与讨论无关，不符合三个维度。

(D) 新闻谈话节目是形成国家政治辩论的主要机制。与是否是主要辩论机制无关。

(E) 正确。新闻谈话节目的主持人和嘉宾更有可能提出与他们自己性别有关的问题。本选项直接指出嘉宾的性别和讨论的事务是相关的。

第二组

1. Allegretto Winery is a small winery whose wines are sought after by collectors. At the end of this year's grape-growing season, there was a long hot dry spell, which kept the grapes small. Since Allegretto uses only grapes grown in its own vineyard, it expects to produce significantly less wine than usual. Nevertheless, Allegretto expects that its revenue from this year's wine will equal its revenue from more normal years.

Which of the following, if true, does most to justify the winery's expectation?

(A) Small grapes do not require significantly less labor to harvest than normal-sized grapes do.

(B) Long hot dry spells at the beginning of the grape-growing season are rare, but they can have a devastating effect on a vineyard's yield.

(C) Many wineries make wine from grapes grown in vineyards they do not own, sometimes even from grapes grown in other regions.

(D) When hot dry spells are followed by heavy rains, the rains frequently destroy grape crops.

(E) Grapes that have matured in hot dry weather make significantly better wine than ordinary grapes do.

文段大意：Allegretto 酒厂是一家小酒厂，其生产的葡萄酒受到收藏家的追捧。在今年葡萄种植季节结束时，有一段长时间的炎热干旱期，这使得葡萄个头很小。由于 Allegretto 酒厂只使用自己葡萄园里种植的葡萄，因此预计产量将大大低于平时。尽管如此，Allegretto 酒厂预计今年葡萄酒的收入将与往年持平。

推理：本题要求我们解释“为什么 Allegretto 酒厂预计其今年的葡萄酒收入将与正常年份的收入持平”。

选项分析：

(A) 与正常大小的葡萄相比，收获小葡萄需要的劳动力并不会少很多。劳动力与收入无关。

(B) 在葡萄种植季节开始时，长时间的高温干旱是罕见的，但它们会对葡萄园的产量产生破坏性的影响。此选项无法解释为什么收入持平。

(C) 许多酒厂用不是自己的葡萄园种植的葡萄酿造葡萄酒，有时甚至用其他地区种植的葡萄。与其他地区的葡萄无关。

(D) 炎热的干旱期过后是大雨，大雨经常会摧毁葡萄作物。此选项无法解释为什么收入持平。

(E) 正确。在炎热干燥天气下成熟的葡萄比普通的葡萄能酿造出明显更好的葡萄酒。如果今年的葡萄酒收入真的能与正常年份的收入持平，那么很可能是因为个头小的葡萄酿的酒比较贵（因为前提中讲过酿的酒少了，那么想收入一样，必然是单价要高）。

2. Which of the following most logically completes the argument below?

The Republic of Naboa is replacing its $2 bills with $2 coins. On average, each coin costs more to produce than does each bill. Nevertheless, producing coins instead of bills will enable the government of Naboa to save substantially on its production costs for $2 currency in future years, since ________.

(A) the coins will be introduced gradually into circulation as worn $2 bills are taken out of circulation

(B) all currency in Naboa of denominations higher than \$2 will continue to be paper currency

(C) consumers are no less likely to use a \$2 coin than they are to use a \$2 bill

(D) the total amount of the metal used per \$2 coin is itself valued at much less than \$2

(E) the coins will remain in circulation for an average of twenty years, whereas \$2 bills generally wear out after only two years

文段大意：纳博亚共和国将用 2 美元硬币取代 2 美元纸币。平均而言，每枚硬币的生产成本高于每张纸币的生产成本。尽管如此，生产硬币而不是纸币将使纳博亚政府在未来几年内大大节省 2 美元货币的生产成本，因为________。

推理：本题要求我们解释“为什么生产硬币而不是纸币将使纳博亚政府在未来几年内大大节省 2 美元货币的生产成本”。

选项分析：

(A) 硬币将随着破旧的 2 美元纸币退出流通而逐渐进入流通领域。与替换的手段无关。

(B) 纳博亚所有面值高于 2 美元的货币将继续使用纸币。与面值高于 2 美元的货币无关。

(C) 消费者使用 2 美元硬币的可能性并不比使用 2 美元纸币的可能性小。无法解释生产成本的问题。

(D) 每枚 2 美元硬币所使用的金属总量本身的价值远远低于 2 美元。与硬币本身的成本无关。需要比较其和纸币的生产成本。

(E) 正确。硬币平均可以流通 20 年，而 2 美元纸币一般在两年后就会磨损。只有本选项解释了为什么生产硬币的“总成本”要比生产纸币更低。

3. Over the past several years coffee consumption in the United States has declined while the total combined revenues of American coffee companies have increased substantially. Although all coffee companies have increased prices

on all brands of coffee they sell, the increase in revenues is much higher than can be accounted for by the increase in prices.

Which of the following, if true, most contributes to an explanation of the increase in revenues not accounted for by the increase in prices?

(A) The decline in coffee consumption has forced many smaller coffee companies to sell their operations to companies that are market leaders.

(B) Many former coffee drinkers have switched to drinking herbal teas and caffeine-free sodas.

(C) Many coffee drinkers have switched from bargain brands of coffee to more expensive premium brands.

(D) Coffee-exporting countries have placed larger export tariffs on coffee to compensate for declining sales.

(E) New brewing technologies allow coffee drinkers to use less ground coffee per cup of brewed coffee.

文段大意：在过去的几年里，美国的咖啡消费量下降了，而美国咖啡公司的总收入却大幅增加。尽管所有咖啡公司都提高了他们销售的所有品牌的咖啡的价格，但收入的增加远远高于价格上涨所能解释的。

推理：本题要求我们解释“为什么收入的增加不是由价格的增加所引起的”。实际上，本题就是要我们找出另一个能导致收入增加的原因。

选项分析：

(A) 咖啡消费量的下降迫使许多较小的咖啡公司将他们的业务卖给了是市场领导者的公司。

(B) 许多以前喝咖啡的人已经转而喝草本茶和不含咖啡因的苏打水。

(C) 正确。许多喝咖啡的人已经从廉价品牌的咖啡转向更昂贵的高级品牌。本选项解释了为什么收入会增加，那是因为顾客买咖啡花费的钱增多了。

(D) 咖啡出口国对咖啡征收更多的出口关税，以弥补销量的下降。

(E) 新的冲泡技术使咖啡饮用者在冲泡的每杯咖啡中使用更少的咖啡粉。

4. Which of the following most logically completes the argument?

The Bluewater Hotel usually earns half its yearly revenue during the year's first quarter, when typically all its rooms are occupied. At other times of the year it has many vacancies. This year the Bluewater was closed throughout the first quarter while damage caused by a severe storm at the end of last year was repaired. Nevertheless, the Bluewater is unlikely to see this year's revenue drop to half of a normal year's revenue, since _______.

(A) the closure this year will not affect its revenue next year, provided the Bluewater is open during next year's first quarter

(B) few of the other hotels in the region were forced to close for repairs after the storm

(C) the cost of the repairs was less than half of a normal year's revenue

(D) it is extremely unlikely that any storm this year will be so severe as to cause damage comparable to the damage caused by the storm last year

(E) many of the customers who would ordinarily have stayed at the Bluewater during the first quarter will probably reschedule their stay for later in the year

文段大意：蓝水酒店（Bluewater Hotel）的年收入通常有一半是在第一季度实现的，一般在这个时候，所有的房间都住满了。在一年中的其他时间，它有很多空置房间。今年，蓝水酒店在第一季度停业，同时去年年底一场强风暴对酒店造成损坏，需对其进行修复。然而，蓝水酒店今年的收入不太可能下降到正常年度收入的一半，因为________。

推理：本题要求我们解释“为什么蓝水酒店今年的收入不太可能下降到正常年份的一半”。

选项分析：

(A) 今年的停业不会影响其明年的收入，只要蓝水酒店明年第一季度开放。与明年的收入无关。

（B）该地区的其他酒店在风暴后很少被迫停业进行维修。与其他酒店无关。

（C）维修费用不到正常年份收入的一半。与成本无关。

（D）今年的任何风暴都极不可能像去年的风暴那么强，造成的损坏也极不可能与去年的风暴相媲美。与去年的风暴无关。

（E）正确。许多通常会在第一季度入住蓝水酒店的客户可能会将他们的入住时间重新安排到今年晚些时候。本选项解释了为什么总收入不会下降，这是因为反正客户早晚都要把钱花在这里。

5. The country of Karlandia imposes a tax of three cents per gallon on sales of gasoline. The revenues from this tax are used to maintain Westland's roads. The number of miles driven per year in Karlandia has doubled since the tax went into effect fifteen years ago, but although taxes are collected as diligently as ever, gasoline tax revenues have not doubled.

Which of the following, if true, best reconciles the apparent discrepancy?

（A）Many miles of new roads, all of which will require regular maintenance, are under construction in Westland.

（B）Overall, vehicles in Karlandia consume less fuel per mile driven than they did 15 years ago.

（C）Gasoline distributors in Karlandia import more gasoline than they did fifteen years ago.

（D）The number of vehicles in Karlandia has increased dramatically in the last fifteen years.

（E）The price of gasoline has not doubled in the last fifteen years.

文段大意：卡兰迪亚对每加仑汽油征收 3 美分的税。这笔税收用于维护韦斯特兰的道路。自从 15 年前汽油税生效以来，卡兰迪亚每年的行驶里程数翻了一番。但是，尽管税收征收一如既往地频繁，汽油税收却没有翻倍。

推理，本题要求我们解释"为什么行驶的里程数增加但汽油税收却没有增加"。

选项分析：

（A）西部地区正在修建数英里长的新道路，这些道路都需要定期维护。与维修道路无关。

（B）正确。总的来说，卡兰迪亚的车辆每行驶一英里所消耗的燃料比 15 年前要少。由于税是按照耗油量来收的，所以即便里程数增加了，政府收到的税不会同比例增加。

（C）卡兰迪亚的汽油经销商比 15 年前进口更多的汽油。与汽油进口量无关。

（D）在过去的 15 年里，卡兰迪亚的车辆数量急剧增加。与汽油税收最直接相关的是汽油，与汽车数量无关。

（E）汽油的价格在过去 15 年里没有翻倍。汽油价格没有翻倍无法合理解释"为什么行驶的里程数增加但汽油税收却没有增加"。

6. Sales of Paradoze, a nonprescription sleeping pill, declined throughout the six months in which a certain television advertisement was used to promote the product. Yet the makers of Paradoze consider the television advertisement to have been very effective.

Which of the following, if true, most helps to resolve the apparent discrepancy?

(A) The makers of Paradoze had previously used television advertising to promote nonprescription products.

(B) On average, only one out of four advertisements for a nonprescription drug is judged effective by the drug's manufacturer.

(C) During the six months in question, Paradoze's share of a shrinking market for nonprescription sleeping pills increased.

(D) Paradoze was the only nonprescription sleeping pill being promoted through television advertisements during the six months in question.

(E) Paradoze's share of the market for nonprescription sleeping pills has always been small in comparison with that of its largest competitor.

文段大意：非处方安眠药 Paradoze 的销量在某电视广告宣传该产品的 6 个月期间有所下降。然而，Paradoze 的制造商却认为电视广告非常有效。

推理：本题要求我们解释“为什么 Paradoze 销量下降但其广告是有效的”。

选项分析：

(A) Paradoze 的制造商以前曾使用电视广告来推广非处方产品。与制造商之前做了什么无关。

(B) 平均来说，非处方药的四则广告中只有一则被药物的制造商认为是有效的。无法合理解释“药物销量下降”与“电视广告有效”之间的关系。

(C) 正确。在有关的六个月内，Paradoze 在不断缩小的非处方安眠药市场上的份额增加了。本选项刚好解释了差异，即因为总市场缩小了，所以药物销量低，但其市场份额的增加足以证明广告是有效的。

(D) Paradoze 是唯一的非处方安眠药，在有关的六个月内通过电视广告进行推广。不管是不是唯一的非处方安眠药，都无法合理解释“药物销量下降”与“电视广告有效”之间的关系。

(E) 与其最大的竞争者相比，Paradoze 在非处方安眠药市场上的份额一直很小。无法合理解释“药物销量下降”与“电视广告有效”之间的关系。

7. In Kravonia, the average salary for jobs requiring a college degree has always been higher than the average salary for jobs that do not require a degree. Current enrollments in Kravonia's colleges indicate that over the next four years the percentage of the Kravonian workforce with college degrees will increase dramatically. Therefore, the average salary for all workers in Kravonia is likely to increase over the next four years.

Which of the following is an assumption on which the argument depends?

(A) Kravonians with more than one college degree earn more, on average, than do Kravonians with only one college degree.

(B) The percentage of Kravonians who attend college in order to earn higher salaries is higher now than it was several years ago.

(C) The higher average salary for jobs requiring a college degree is not due largely to a scarcity among the Kravonian workforce of people with a college degree.

(D) The average salary in Kravonia for jobs that do not require a college degree will not increase over the next four years.

(E) Few members of the Kravonian workforce earned their degrees in other countries.

类别：普通预测推理

前提：在接下来的四年里，拥有大学学位的劳动力比例将大幅增加。

结论：所有工人的平均工资在未来四年里可能会增加。

推理：结论“平均工资在未来四年里可能会增加”的充分条件应为：未来四年，大学生的比例增加且工资配比模式和之前相同。（原文明确给出了到现在为止，大学生的工资是高的）。因此，答案选项必须指出“未来四年的工资配比模式和之前相同”。

选项分析：

(A) 平均来说，有超过一个学位的 Kravonia 人比只有一个学位的 Kravonia 人挣得会稍微多一点。本选项提到了“薪资”，但是并不是将有学位的和没有学位的 Kravonia 人相比较。

(B) 为了挣得高薪而进入大学学习的 Kravonia 人比以前多了一些。本选项讨论的是拥有“大学文凭”的目的。

(C) 正确。要求大学学位的工作岗位薪资高并不是因为拥有大学学位的 Kravonia 人少而导致的。本选项直接修正了偏差，即这些有文凭的人可以获得那些高薪的职位，而不会因为有文凭的人变多了而导致高薪工作机会减少。

(D) Kravonia 不需要大学学位的工作的平均工资在未来四年内不会增加。本选项和结论无关。

(E) 很少有 Kravonia 人在别的国家拿到文凭。本选项和结论无关。

8. Until now, only injectable vaccines against influenza have been available. Parents are reluctant to subject children to the pain of injections, but adults, who are at risk of serious complications from influenza, are commonly vaccinated. A new influenza vaccine, administered painlessly in a nasal spray, is effective for children. However, since children seldom develop serious complications from influenza, no significant public health benefit would result from widespread vaccination of children using the nasal spray.

Which of the following is an assumption on which the argument depends?

(A) Any person who has received the injectable vaccine can safely receive the nasal-spray vaccine as well.

(B) The new vaccine uses the same mechanism to ward off influenza as injectable vaccines do.

(C) The injectable vaccine is affordable for all adults.

(D) Adults do not contract influenza primarily from children who have influenza.

(E) The nasal spray vaccine is not effective when administered to adults.

类别：普通预测推理

前提：儿童很少出现严重的流感并发症。

结论：广泛使用鼻腔喷雾剂接种儿童疫苗不会对公共卫生产生显著的益处。

推理：结论“广泛使用鼻腔喷雾剂接种儿童疫苗不会对公共卫生产生显著的益处”的充分条件应为：儿童感染流感对于整体公共卫生来说影响不大。因此，答案选项必须指出“儿童不会出现严重的流感并发症等于不会影响整体公共卫生”。

选项分析：

(A) 任何已经接种疫苗的人也可以安全地接种鼻喷雾疫苗。与已经接种疫苗的人能否接种鼻喷雾疫苗无关。

(B) 新型疫苗和注射型疫苗抵抗流感的原理相同。抗病原理和是否有益于公共卫生无关。

(C) 注射型疫苗是所有成人都能负担的。与接种疫苗的价格无关。

(D) 正确。成人的流感并发症并非是孩子传染的。如果成人的流感是孩子传染的，那么孩子接受鼻喷雾疫苗，即使对自身没有什么益处，但可以对成人有益，成人也是“整个公共卫生”的一部分。所以，我们需要排除这种情况，确保儿童可以代表整个公共卫生。

(E) 鼻喷雾疫苗对成人没有作用。因为成人可以接种注射型疫苗，所以鼻喷雾疫苗对成人是否有效果与结论没有关系。

9. Newspaper editorial: A surprising study by the World Health Organization found that chronic exposure to excessive noise may cause or exacerbate serious health problems such as stroke and heart disease. This study will produce health benefits for our city, for it will inevitably lead to pressure on our local politicians, who will respond by passing stricter noise ordinances.

The editorial's reasoning relies on which of the following assumptions?

(A) The most effective way to reduce the public's exposure to excessive noise is through stricter noise ordinances.

(B) Stricter noise ordinances will lessen the amount of excessive noise people would otherwise be exposed to.

(C) Rates of heart disease and stroke tend to be highest in areas where excessive noise is great.

(D) Local politicians who help pass stricter noise ordinances will do so primarily to improve the health of their constituents.

(E) The newspaper's city currently has less strict noise ordinances than do many other cities of comparable size.

类别：普通预测推理

前提：某项研究迫使政治家将通过更严格的噪声条例。

结论：该研究将对我们的城市产生健康益处。

推理：结论"该研究将对我们的城市产生健康益处"的充分条件应为：某项研究迫使政治家将通过更严格的噪声条例，且公众会遵守该条例。因此，答案选项必须指出"公众会遵守该条例"。

选项分析：

(A) 使公众接触到的过量噪声减少的最有效方法是通过更严格的噪声条例。此选项容易误选。注意：强调"最有效"是没有意义的，可能其他方法更差，"通过噪音条例"这个方法也只是矮子里选将军，不代表此方法本身有效。
(B) 正确。更严格的噪声条例将使人们接触到的过量噪声减少。
(C) 在噪声过大的地区，心脏病和中风的发病率往往最高。
(D) 帮助通过更严格的噪声条例的当地政治家将这样做，主要是为了改善他们选民的健康。
(E) 该报的所在城市目前的噪声条例没有其他许多同等规模的城市那么严格。

10. For similar cars and drivers, automobile insurance for collision damage has always cost more in Greatport than in Fairmont. Police studies, however, show that cars owned by Greatport residents are, on average, slightly less likely to be involved in a collision than cars in Fairmont. Clearly, therefore, insurance companies are making a greater profit on collision-damage insurance in Greatport than in Fairmont.

Which of the following is an assumption on which the argument depends?

(A) Repairing typical collision damage does not cost more in Greatport than in Fairmont.

(B) There are no more motorists in Greatport than in Fairmont.

(C) Greatport residents who have been in a collision are more likely to report it to their insurance company than Fairmont residents are.

(D) Fairmont and Greatport are the cities with the highest collision-damage insurance rates.

(E) The insurance companies were already aware of the difference in the likelihood of collisions before the publication of the police reports.

类别：普通预测推理

前提：G 国的保费比 F 国高；G 国的事故率比 F 国低。

结论：G 国比 F 国的保险利润更高。

推理：结论"G 国比 F 国的保险利润更高"的充分条件应为：收入 – 成本的差值更高。因此，答案选项必须指出总体成本的情况。

选项分析：

(A) 正确。在 Greatport 修理典型的碰撞损坏不会比 Fairmont 贵。对于保险公司来说，保费可以看作收入，但事故率并不能完全等于成本，修理被撞坏的汽车也是成本之一。

(B) Greatport 的汽车驾驶员的数量不比 Fairmont 的汽车驾驶员数量多。由于保险公司的成本仅仅和出事故的车辆有关系，所以汽车总数或者说驾驶员总数不会影响结论的产生。

(C) 发生碰撞事故的 Greatport 居民比 Fairmont 居民更有可能向保险公司汇报自己出了状况。

(D) Greatport 和 Fairmont 是两个最容易发生碰撞事故的城市。本选项描述了两个城市的情况，不涉及结论中的利润内容。

(E) 在警察公布调查结果之前，保险公司就已经知道发生碰撞事故可能性的差异。本选项和利润无关。

2.3 ▸ 评估论证 （Critique）

评估论证类考题主要考查我们对于“假说论证”的深刻理解。请大家在看本节内容之前，先复习一下第一章中关于“假说论证”的全部内容。

评估论证类考题的问法可分为削弱类、加强类和评价类。

削弱类问题包括但不仅限于：

（1）Which of the following, if true, most seriously weakens the conclusion above?

（2）Which of the following statements, if true, would cast the most doubt on the conclusion drawn above?

（3）Which of the following, if true, most seriously calls Scott's hypothesis into question?

（4）Which of the following, if true, most seriously undermines the conclusion drawn above?

（5）The banker's argument is most vulnerable to criticism on which of the following grounds?

（6）Which of the following points to the most serious logical flaw in the reviewer's argument?

加强类问题包括但不仅限于：

（1）Which of the following, if true, most strengthens the argument above?

（2）Which of the following, if true, most supports the researchers' conclusion?

（3）Which of the following, if true, provides the strongest grounds for the experts' conclusion?

评估类问题包括但不仅限于：

Which of the following would be most useful to establish in order to evaluate the analyst's prediction?

这三类问题从本质上来说是没有区别的。例如：

> 小明扶着老奶奶过马路。因此，小明是乐于助人的人。
>
> 以下哪个选项，如果正确，能最强地削弱上述推理？
>
> (A) 除非许以回报，否则小明从不对别人伸出援手。
> (B) 是否乐于助人是衡量一个孩子好坏的重要标准。

例题答案为选项A。这道题可以改为：

> 小明扶着老奶奶过马路。因此，小明是乐于助人的人。
>
> 以下哪个选项，如果正确，能最强地加强上述推理？
>
> (A) 在不被许以回报的情况下，小明依然经常对别人伸出援手。
> (B) 是否乐于助人是衡量一个孩子好坏的重要标准。

答案依然为选项A，这道题还可以改为：

> 小明扶着老奶奶过马路。因此，小明是乐于助人的人。
>
> 以下哪个选项，如果正确，能评价上述推理？
>
> (A) 是否是在被许以回报后，小明才对别人伸出援手？
> (B) 是否乐于助人是衡量一个孩子好坏的重要标准吗？

答案还是选项A。

五种假说论证

我们在第一章讲过，假说论证总共分为五种：

(1) 普通预测推理

(2) 泛化推理

(3) 类比推理

(4) 归因推理

(5) 因果推理

我们需要充分掌握每个类型的特点，准确识别后解题。

2.3.1 普通预测推理

这类推理在 GMAT 考试中是十分常见的。它们的前提是一个现象或原理，结论则是作者基于这个现象或原理对未来做出的预判或对其可能产生的结果的叙述。

例如：

前提：A 公司成本增加。

结论：该公司的利润将会下降。

前提“A 公司成本增加”是一个现象，结论“该公司的利润将会下降”是基于该现象的对未来将要发生的事情的一个预测。

这类推理的漏洞在于前提并非结论成立的“充分条件”，而仅是“其中之一”的条件或仅“与结论相关”。因此，想要评估这类推理，我们需首先分析出结论在前提所讲的方向上的充分条件是什么，进而根据前提与该充分条件的差距确定答案方向。

在上例中，利润下降的充分条件应是"A 公司的总收入和总成本的差值下降"。因为前提中只讲到了成本增加，所以答案选项必然要说到 A 公司总收入的情况。

又例如：

> 前提：有人要进入小明家。
>
> 结论：小明家的门锁一定会被撬开。

"门锁被撬开"的充分条件应是"有人要强行闯入小明家"。因为前提中只讲到了有人要进入小明家，所以答案选项必然要考虑前提讲的"进入"是否是"强行闯入"。

综上所述，普通预测推理类题的解题步骤如下：

> （1）确认前提和结论
>
> （2）分析结论在前提所讲的方向上的充分条件
>
> （3）根据前提与该充分条件的差距确定答案方向
>
> （4）选出答案

例题 1

Many people suffer an allergic reaction to sulfites, including those that are commonly added to wine as preservatives. However, since there are several winemakers producing wines to which no sulfites are added, those who would like to drink wine but are allergic to sulfites can drink these wines without risking an allergic reaction to sulfites.

Which of the following, if true, most seriously weakens the argument?

(A) Sulfites occur naturally in most wines.

(B) The sulfites that can produce an allergic reaction are also commonly found in beverages other than wines.

(C) Wines without added sulfites tend to be at least moderately expensive.

(D) Apart from sulfites, there are other substances commonly present in wine that can trigger allergic reactions.

(E) Wine without added sulfites sometimes becomes undrinkable even before the wine is sold to consumers.

(1) 上述例题的前提和结论可以整理为：

前提：酒里不添加亚硫酸盐。

结论：那些对亚硫酸盐过敏的人可以不用冒着对亚硫酸盐过敏的风险喝酒了。

(2) 结论"不对亚硫酸盐过敏"的充分条件应为"酒里不含亚硫酸盐"。

(3) 前提只讲到了没有亚硫酸盐的某一方面，所以答案选项应给出可能存在其他的亚硫酸盐来源。

(4) 显然选项 A 是正确的（选项 D 不正确是因为结论讲的是对亚硫酸盐过敏，因此对其他物质过敏和推理无关）。它给出了另外一些亚硫酸盐的来源。

例题 2

Lightbox, Inc., owns almost all of the movie theaters in Washington County and has announced plans to double the number of movie screens it has in the county within five years. Yet attendance at Lightbox's theaters is only just large enough for profitability now and the county's population is not expected to increase over the next ten years. Clearly, therefore, if there is indeed no increase in population, Lightbox's new screens are unlikely to prove profitable.

Which of the following, if true about Washington County, most seriously weakens the argument?

(A) Though little change in the size of the population is expected, a pronounced shift toward a younger, more affluent, and more entertainment-oriented population is expected to occur.

(B) The sales of snacks and drinks in its movie theaters account for more of Lightbox's profits than ticket sales do.

(C) In selecting the mix of movies shown at its theaters, Lightbox's policy is to avoid those that appeal to only a small segment of the moviegoing population.

(D) Spending on video purchases, as well as spending on video rentals, is currently no longer increasing.

(E) There are no population centers in the county that are not already served by at least one of the movie theaters that Lightbox owns and operates.

(1) 上述例题的前提和结论可以整理为：

前提：华盛顿县的人口在未来十年内预计不会增加。

结论：Lightbox 电影院新增的电影屏幕赚不到钱。

(2) 结论“Lightbox 电影院新增的电影屏幕赚不到钱”的充分条件应为“没有更多人来看电影”。

(3) 前提只讲到了人口不会增多，所以答案选项应给出“人口”和“看电影的人”不相同。

选项分析

(A) 正确。虽然预计人口规模变化不大，但预计人口将明显转向更年轻、更富裕、更注重娱乐的人群。此选项直接指出“华盛顿县的人口不增加≠来看电影的人不增加”。

（B）在 Lightbox 电影院中，零食和饮料的销售比电影票的销售给其带来更多利润。零食、饮料的销售利润“占比高”不等于本身卖得多，占比和本身的销售额不能画等号。

（C）在选择电影院放映的电影组合时，Lightbox 的政策是避免那些只吸引一小部分观影人群的电影。选片政策和讨论无关。

（D）购买视频和租赁视频的支出目前都不再增加。电影院的利润与视频的购买或租赁无关。

（E）Lightbox 拥有并经营的至少一家电影院已经为该县的人口中心提供服务。与服务对象无关。

例题 3

Drug manufacturer: We have just discovered that Rodol, one of our drugs, kills the bacteria that cause tullis disease, a previously untreatable chronic ailment. Although Rodol has harmful side effects for people who have severe liver damage, a recent examination of hospital records shows that only one person in ten thousand has severe liver damage. So Rodol can safely be prescribed to most tullis disease patients.

Which of the following would be most useful in evaluating the drug manufacturer's argument?

(A) Whether there are more than ten thousand tullis disease patients in the market served by the drug manufacturer

(B) Whether the symptoms of tullis disease are similar to the symptoms of any diseases for which Rodol would be unhelpful

(C) Whether liver disease is more common among people with tullis disease than it is for the population in general

(D) How long a course of treatment with Rodol is required for a tullis disease patient to recover fully

(E) For which diseases other than tullis disease Rodol is an effective and safe treatment

(1) 上述例题的前提和结论可以整理为：

前提：每一万人中只有一个人有严重的肝损伤（Rodol 对有肝损伤的人有副作用）。

结论：大多数 tullis 患者可以安全地服用 Rodol。

(2) 结论"大多数 tullis 患者可以安全地服用 Rodol"的充分条件应为"tullis 患者服用后没有副作用"。

(3) 前提只讲到了普通人大部分服用后没有副作用，所以答案选项应考虑"普通人"和"tullis 患者"在服药后的副作用方面是否不同。

选项分析

(A) 在该药品制造商所服务的市场上是否有超过一万名 tullis 患者。与 tullis 患者本身有多少无关。

(B) tullis 病的症状是否与 Rodol 无效的任何疾病的症状相似。与其他疾病无关。

(C) 正确。肝脏疾病在 tullis 患者中是否比在普通人群中更常见。

(D) tullis 患者需要多长时间的 Rodol 治疗才能完全康复。与康复时长无关。

(E) 除了 tullis 病外，Rodol 对哪些疾病是有效和安全的治疗方法。与 Rodol 还能治疗什么疾病无关。

2.3.2 泛化推理

泛化推理通常使用样本（前提）来支持总体（结论）。在 GMAT 考题中，绝大部分泛化推理的前提会给出明显的调查数据，结论对该数据进行归纳总结。

评价一个泛化推理的好坏主要有两个维度：

（1）样本是否存在偏差

（2）样本数量是否足够

泛化推理类题的解题步骤如下：

（1）确认前提和结论

（2）确定样本和总体

（3）思考样本是否存在偏差以及数量是否足够

（4）选出答案

例题 1

Business analyst：National Motors began selling the Luxora—its new model of sedan—in June. Last week，National released sales figures for the summer months of June，July，and August that showed that by the end of August only 80, 000 Luxoras had been sold. Therefore，National will probably not meet its target of selling 500, 000 Luxoras in the model's first twelve months.

Which of the following would be most useful to establish in order to evaluate the analyst's prediction?

（A）Whether new car sales are typically lower in the summer months than at any other time of the year

（B）Whether National Motors currently produces more cars than any other automaker

(C) Whether the Luxora is significantly more expensive than other models produced by National Motors

(D) Whether National Motors has introduced a new model in June in any previous year

(E) Whether National Motors will suffer serious financial losses if it fails to meet its sales goal for the Luxora

(1) 上述例题的前提和结论可以整理为：

前提：National Motors 在六月、七月和八月这三个月内总共卖了 8 万辆 Luxoras。

结论：National Motors 在前 12 个月内无法卖超过 50 万辆 Luxoras。

(2) 样本为六月、七月和八月的 Luxoras 销量。

(3) 考虑六月、七月和八月这三个月是否是销售淡季或者考虑是否能给出更多月份的销售数据。

(4) 显然，只有选项 A（正常来说，新车的销量在夏季月份是否都不理想）在考虑样本是否有偏差，所以答案为 A。

例题 2

Many in the medical profession are concerned that the widespread use of growth-promoting antibiotics in animal feed is reducing the antibiotics' effectiveness in humans, because microbes can develop resistance to the drugs. Now MagnaBurger, a large fast-food chain, is considering agreeing not to buy meat from producers that use antibiotics as growth promoters. Such an agreement would certainly induce many producers to stop the use of antibiotics as growth promoters. Thus, if MagnaBurger decides to do this, it will probably reduce significantly the extent to which antibiotics are used as growth promoters.

Which of the following would, if true, most strengthen the argument above?

(A) It is considerably more expensive to raise animals without using growth-promoting antibiotics.

(B) Some countries have banned the general use of antibiotics in animal feed.

(C) If MagnaBurger stops buying meat from producers that use antibiotics as growth promoters, other large fast-food chains are likely to do so, too.

(D) If MagnaBurger is the only fast-food chain that stops buying meat antibiotics, the prices that its competitors pay to purchase meat will decline.

(E) One study showed that there would be a dramatic drop in the incidence of antibiotic-resistant bacteria if antibiotics were no longer used as growth promoters in agriculture.

(1) 上述例题的前提和结论可以整理为:

前提:麦格纳汉堡公司的行为肯定会促使许多生产商停止使用抗生素作为生长促进剂。

结论:麦格纳汉堡公司的做法会大大减少抗生素作为生长促进剂的使用范围。

(2) 样本为许多生产商。

(3) 说明前提讲到的生产商能代表所有使用抗生素作为生长促进剂的生产商或者说明更多生产商有类似行为。

(4) 显然,本题答案为选项C(如果麦格纳汉堡公司停止从使用抗生素的生产者那里购买肉类,那么其他大型快餐连锁店也可能这样做),它给出了更多符合结论说法的样本,因此加强了结论。

2.3.3 类比推理

类比推理首先提到两个或两个以上的事物在某些方面是相似的，然后将关于一个事物的主张作为接受关于另一个事物的相似主张的理由。

在评估一个类比推理时，需要考虑是否存在“其他相关相似点缺失（the absence of additional relevant similarities）”。

类比推理类题的解题步骤如下：

（1）确认前提和结论
（2）确定类比对象
（3）思考类比对象是否存在相关相似点缺失
（4）选出答案

例题

Because visual inspection cannot reliably distinguish certain skin discolorations from skin cancers, dermatologists at clinics have needed to perform tests of skin tissue taken from patients. At Westville Hospital, dermatological diagnostic costs were reduced by the purchase of a new imaging machine that diagnoses skin cancer in such cases as reliably as the tissue tests do. Consequently, even though the machine is expensive, a dermatological clinic in Westville is considering buying one to reduce diagnostic costs.

Which of the following would it be most useful for the clinic to establish in order to make its decision?

(A) Whether the visits of patients who require diagnosis of skin discolorations tend to be shorter in duration at the clinic than at the hospital

(B) Whether the principles on which the machine operates have been known to science for a long time

(C) Whether the machine at the clinic would get significantly less heavy use than the machine at the hospital does

(D) Whether in certain cases of skin discoloration, visual inspection is sufficient to make a diagnosis of skin cancer

(E) Whether hospitals in other parts of the country have purchased such imaging machines

(1) 上述例题的前提和结论可以整理为:

前提:在韦斯特维尔医院,由于购买了一台新的成像机器,皮肤病诊断费用减少了。

结论:韦斯特维尔的一家皮肤病诊所正在考虑购买一台成像机器,以降低诊断成本。

(2) 类比对象为“韦斯特维尔医院”和“韦斯特维尔的皮肤病诊所”。

(3) “韦斯特维尔医院”和“韦斯特维尔的皮肤病诊所”在使用机器减少成本方面可能存在区别。

(4) 选项A和C都讲到了类比对象的区别,但和使用机器减少成本方面更为相关的是C,即机器使用的频繁度的区别。

2.3.4 归因推理

归因推理也是一种常见的假说论证方式,其核心在于探究前提中现象的成因,或寻找解释该现象发生的原理(动机)。例如:

前提：家里被翻找得十分杂乱。

结论：可能是有小偷来过了。

在这个例子中，结论在探究“家里为何会很杂乱”，所以是典型的归因推理。评估归因推理有两个维度：

（1）是否有其他证据证明结论成立

（2）是否有其他可能的原因和解释

归因推理类题的解题步骤如下：

（1）确认前提和结论

（2）思考是否存在令结论成立的其他证据以及前提是否存在其他可能的解释

（3）选出答案

例题 1

Paleontologists in 1947 found a large group of fossilized skeletons of the carnivorous dinosaur Coelophysis. In the space where one Coelophysis's stomach had been, there were bones from another individual of the same species. The paleontologists concluded from this that Coelophysis was cannibalistic.

In order to evaluate the strength of the paleontologists' argument, it would be most helpful to know which of the following?

(A) How many other dinosaur species were also found mixed among the Coelophysis

(B) To what extent bones from different individual Coelophysis dinosaurs were randomly jumbled together in the group of fossilized skeletons

(C) What bones of other dinosaur species were also found mixed among the Coelophysis bones

(D) How many other dinosaur species Coelophysis is known to have eaten

(E) Whether the bones that were found in the space that had been one Coelophysis's stomach came from more than one part of the other Coelophysis's body

(1) 上述例题的前提和结论可以整理为：

前提：在一只 Coelophysis 的胃空间里，有同一物种另一个体的骨头。

结论：Coelophysis 是自食的（自己吃自己）。

(2)“恐龙自食”这个习性是否存在更多的其他证据，例如，是否有自食的基因等；是否存在其他可能的解释，可说明为什么“在 Coelophysis 的胃空间里，有同一物种另一个体的骨头”。

选项分析

(A) 在 Coelophysis 中还发现有多少其他的恐龙物种混合在一起。

(B) 正确。在多大程度上，来自不同个体的 Coelophysis 恐龙的骨头随机混杂在了这组骨骼化石中？显然，如果本选项回答“是”，则它是一个可以解释为什么“在 Coelophysis 的胃空间里，有同一物种另一个体的骨头”的原因。

(C) 哪些其他种类的恐龙的骨头也被发现混杂在 Coelophysis 的骨头中。

(D) 已知 Coelophysis 共吃过多少种其他恐龙。

(E) 在曾是一只 Coelophysis 的胃的空间里发现的骨头是否来自另一只 Coelophysis 的不止一个部位。

例题 2

Every fall Croton's jays migrate south. The jays always join flocks of migrating crookbeaks with which they share the same summer and winter territories. If a jay becomes separated from the crookbeaks it is accompanying, it wanders until it comes across another flock of crookbeaks. Clearly, therefore, Croton's jays lack the navigational ability to find their way south on their own.

Which of the following, if true, most strengthens the argument above?

(A) Croton's jays lay their eggs in the nests of crookbeaks, which breed upon completing their southern migration.

(B) The three species most closely related to crookbeaks do not migrate at all.

(C) In the spring, Croton's jays migrate north in the company of Tattersall warblers.

(D) Species other than Croton's jays occasionally accompany flocks of migrating crookbeaks.

(E) In the spring, crookbeaks migrate north before Croton's jays do.

(1) 上述例题的前提和结论可以整理为：

前提：jay 鸟和 crookbeaks 会一起迁徙，且如果 jay 鸟掉队，会等下一波 crookbeaks 过来再一起走。

结论：jay 鸟缺乏导航能力。

(2) “jay 鸟缺乏导航能力”存在更多的其他证据，例如，jay 鸟的大脑中缺少某种用于导航的器官；不存在其他可以解释为什么“jay 鸟和 crookbeaks 会一起迁徙”的原因，例如，jay 鸟不是因为需要在 crookbeaks 的巢中下蛋才和 crookbeaks 一起行动的。

选项分析

（A）jay 鸟会在 crookbeaks 的鸟巢中下蛋，crookbeaks 会在向南迁徙的过程中孵化这些蛋。本选项可以解释为什么 jay 鸟会跟着 crookbeaks 前行，即是因为要在 crookbeaks 这种鸟的巢里下蛋，所以才会选择跟着 crookbeaks 迁徙，可以削弱推理文段。

（B）和 crookbeaks 最相近的三个种类的鸟从来不迁徙。本选项不符合任何一个维度，可以排除。

（C）正确。在春季，jay 鸟会随着塔特萨尔莺的队伍向北迁徙。本选项给出了 jay 鸟没有方向感的另一个证据。

（D）除了 jay 鸟，其他种类的鸟很少和 crookbeaks 一起成群迁徙。本选项和结论无关。

（E）在春季，crookbeaks 会先于 jay 鸟向北迁徙。jay 鸟不跟着 crookbeaks 不代表它不跟着其他鸟走，因此本选项无法评估推理。

2.3.5 因果推理

凡是前提展示两个事件同时或先后发生并且结论展示了这两个事件的因果关系的推理都是因果推理。

评估因果推理有三个维度：

（1）纯粹巧合

（2）因果倒置

（3）他因导致结果

因果推理类题的解题步骤如下：

(1) 确认前提和结论

(2) 确认结论中的“因”和“果”

(3) 思考三个评估维度

(4) 选出答案

例题 1

Science journalist: A study led by Dr. Elaine Hardman shows that consuming walnuts reduces the risk of breast cancer in at least some types of mice. The study compared the effects of a typical diet and a diet containing walnuts. During the study period, the group whose diet included walnuts developed breast cancer at less than half the rate of the typical-diet group. In addition, the number and sizes of tumors were significantly smaller.

Which of the following would, if true, most undermine the journalist's reasoning?

(A) The mice used in the study received the walnuts directly as food and indirectly through their mothers when they were in utero and while they were nursing.

(B) The amount of walnuts in the daily test diet equates to about two human servings of walnuts.

(C) Walnuts contain multiple ingredients that, individually, have been shown in other studies to reduce the risk of cancer or slow its growth.

(D) In the study's mice, the walnut-containing diet changed the activity of multiple genes that are relevant to breast cancer in both mice and humans.

(E) To keep total dietary fat balanced in the mouse groups, unhealthy fats that lower resistance to disease were reduced by an amount equal to the amount of healthy fats added by the walnuts.

(1) 上述例题的前提和结论可以整理为：

前提：饮食中包含核桃的小组患乳腺癌的比例不到普通饮食小组的一半。

结论：食用核桃至少可以降低某些类型的小鼠患乳腺癌的风险。

显然，本题前提给出了“食用核桃”和“患乳腺癌”呈负相关，结论给出了两者的因果关系，因此属于因果推理。

(2) 因是“食用核桃”；果是“患乳腺癌”。

(3) 思考三个评估维度：

纯粹巧合：指出“食用核桃”和“患乳腺癌”可能根本不具备能产生因果关系的原理；或给出在其他研究中，“食用核桃”和“患乳腺癌”不具备正相关的关系。

因果倒置：指出可能是那些本身就不太可能患乳腺癌的小鼠才喜欢吃核桃。

他因导致结果：给出其他可能导致小鼠患乳腺癌风险低的因素，诸如，那些吃核桃的小鼠本身更喜欢锻炼，而锻炼是降低患乳腺癌风险的重要因素。

(4) 选项E很好地指出了“他因导致结果”，即可能是不健康的脂肪被减少导致小鼠患乳腺癌的风险下降，而不是食用核桃。

例题2

A study followed a group of teenagers who had never smoked and tracked whether they took up smoking and how their mental health changed. After one year, the incidence of depression among those who had taken up smoking was four times as high as it was among those who had not. Since nicotine in cigarettes changes brain chemistry, perhaps thereby affecting mood, it is likely that smoking contributes to depression in teenagers.

Which of the following, if true, most strengthens the argument?

(A) Participants who were depressed at the start of the study were no more likely to be smokers after one year than those who were not depressed.

(B) The study did not distinguish between participants who smoked only occasionally and those who were heavy smokers.

(C) Few, if any, of the participants in the study were friends or relatives of other participants.

(D) Some participants entered and emerged from a period of depression within the year of the study.

(E) The researchers did not track use of alcohol by the teenagers.

(1) 上述例题的前提和结论可以整理为：

前提：凡是“吸烟”的人都更容易“抑郁”（“吸烟”和“抑郁”之间存在正相关关系）。

结论：吸烟导致抑郁。

显然，本题前提给出了“吸烟”和“抑郁”呈正相关，结论给出了两者的因果关系，因此属于因果推理。

(2) 因是“吸烟”；果是“抑郁”。

(3) 思考三个评估维度：

纯粹巧合（加强）：指出“吸烟”和“抑郁”之间存在能产生因果关系的原理；或给出在其他研究中，“吸烟”和“抑郁”依然具备正相关的关系。

因果倒置（加强）：排除那些本身就抑郁的人更可能抽烟。

他因导致结果（加强）：排除其他可能导致抑郁的因素，比如，那些抽烟的人情绪并非一直低迷。

（4）选项 A 很好地指出了“因果倒置（加强）”，即不会是抑郁的人都更容易抽烟。

练 习

第一组

1. The insecticide occurs throughout the plant, including its pollen. Maize pollen is dispersed by the wind and frequently blows onto milkweed plants that grow near maize fields. Caterpillars of monarch butterflies feed exclusively on milkweed leaves. When these caterpillars are fed milkweed leaves dusted with pollen from modified maize plants, they die. Therefore, by using genetically modified maize, farmers put monarch butterflies at risk.

Which of the following would it be most useful to determine in order to evaluate the argument?

(A) Whether the natural insecticide is as effective against maize-eating insects as commercial insecticides typically used on maize are

(B) Whether the pollen of genetically modified maize contains as much insecticide as other parts of these plants

(C) Whether monarch butterfly caterpillars are actively feeding during the part of the growing season when maize is releasing pollen

(D) Whether insects that feed on genetically modified maize plants are likely to be killed by insecticide from the plant's pollen

(E) Whether any maize-eating insects compete with monarch caterpillars for the leaves of milkweed plants growing near maize fields

2. Banking-industry analysts have been predicting the end of the paper check as a method of making payments, now that electronic transfer of funds from one bank account to another is becoming increasingly common. To process paper checks, banks rely on companies that charge a significant amount for each

check they process. Therefore, banks would be well advised to encourage the replacement of paper checks with electronic transfers as the standard method of payment.

Which of the following, if true, most seriously weakens the argument?

(A) Even consumers who currently make use of electronic transfers to pay their regularly recurring bills still commonly use paper checks to make payments for occasional purchases.

(B) Some consumer checking accounts pay the account holder interest on checking account balances.

(C) Government agencies that make regular payments to individuals are increasingly requiring that the people to whom they make such payments receive the funds by electronic transfers into a bank account.

(D) Banks are able to charge their customers substantial fees for many of the services associated with paper-check transactions.

(E) The number of paper-check transactions has decreased only somewhat even though electronic fund transfers have become more common.

3. Because of steep increases in the average price per box of cereal over the last 10 years, overall sales of cereal have recently begun to drop. In an attempt to improve sales, one major cereal manufacturer reduced the wholesale prices of its cereals by 20 percent. Since most other cereal manufacturers have announced that they will follow suit, it is likely that the level of overall sales of cereal will rise significantly.

Which of the following would it be most useful to establish in evaluating the argument?

(A) Whether the high marketing expenses of the highly competitive cereal market led to the increase in cereal prices

(B) Whether cereal manufacturers use marketing techniques that encourage brand loyalty among consumers

(C) Whether the variety of cereals available on the market has significantly increased over the last 10 years

(D) Whether the prices that supermarkets charge for these cereals will reflect the lower prices the supermarkets will be paying the manufacturers

(E) Whether the sales of certain types of cereal have declined disproportionately over the last 10 years

4. Wind farms, which generate electricity using arrays of thousands of wind-powered turbines, require vast expanses of open land. County X and County Y have similar terrain, but the population density of County X is significantly higher than that of County Y. Therefore, a wind farm proposed for one of the two counties should be built in County Y rather than in County X.

Which of the following, if true, most seriously weakens the planner's argument?

(A) County X and County Y are adjacent to each other, and both are located in the windiest area of the state.

(B) The total population of County Y is substantially greater than that of County X.

(C) Some of the electricity generated by wind farms in County Y would be purchased by users outside the county.

(D) Wind farms require more land per unit of electricity generated than does any other type of electrical-generation facility.

(E) Nearly all of County X's population is concentrated in a small part of the county, while County Y's population is spread evenly throughout the county.

5. Because it was long thought that few people would watch lengthy televised political messages, most televised political advertisements, like commercial

advertisements, took the form of short messages. Last year, however, one candidate produced a half-hour-long advertisement. At the beginning of the half-hour slot a substantial portion of the viewing public had tuned into that station. Clearly, then, many more people are interested in lengthy televised political messages than was previously thought.

Which of the following, if true, most seriously weakens the argument?

(A) The candidate who produced the half-hour-long advertisement did not win election at the polls.

(B) The half-hour-long advertisement was widely publicized before it was broadcast.

(C) The half-hour-long advertisement was aired during a time slot normally taken by one of the most popular prime-time shows.

(D) Most short political advertisements are aired during a wide range of programs in order to reach a broad spectrum of viewers.

(E) In general a regular-length television program that features debate about current political issues depends for its appeal on the personal qualities of the program's moderator.

6. Two computer companies, Garnet and Renco, each pay Salcor to provide health insurance for their employees. Because early treatment of high cholesterol can prevent strokes that would otherwise occur several years later, Salcor encourages Garnet employees to have their cholesterol levels tested and to obtain early treatment for high cholesterol. Renco employees generally remain with Renco only for a few years, however. Therefore, Salcor lacks any financial incentive to provide similar encouragement to Renco employees.

Which of the following, if true, most seriously weakens the argument?

(A) Early treatment of high cholesterol does not eliminate the possibility of a stroke later in life.

(B) People often obtain early treatment for high cholesterol on their own.

(C) Garnet hires a significant number of former employees of Renco.

(D) Renco and Garnet have approximately the same number of employees.

(E) Renco employees are not, on average, significantly younger than Garnet employees.

7. In the United States, of the people who moved from one state to another when they retired, the percentage who retired to Florida has decreased by three percentage points over the past ten years. Since many local businesses in Florida cater to retirees, this decline is likely to have a noticeably negative economic effect on these businesses.

Which of the following, if true, most seriously weakens the argument?

(A) Florida attracts more people who move from one state to another when they retire than does any other state.

(B) The number of people who move out of Florida to accept employment in other states has increased over the past ten years.

(C) There are far more local businesses in Florida that cater to tourists than there are local businesses that cater to retirees.

(D) The total number of people who retired and moved to another state for their retirement has increased significantly over the past ten years.

(E) The number of people who left Florida when they retired to live in another state was greater last year than it was ten years ago.

8. Although automobile manufacturers in Brunia charge the same prices for automobiles as do automobile manufacturers in the neighboring country of Corland and the tax structures of the two countries are nearly identical, the hourly wage that Brunian automobile manufacturers pay their employees is

much lower. Therefore, unless automobile manufacturers in Brunia pay substantially more for their raw materials, they make higher profits than do automobile manufacturers in Corland.

Which of the following, if true, most seriously weakens the argument?

(A) Automobile manufacturers in Corland pay more for some raw materials than do automobile manufacturers in Brunia.

(B) In Brunia but not in Corland, automobile manufacturers are required by law to pay their employees health care costs in addition to their wages.

(C) In Corland but not in Brunia, most automobile manufacturers restrict their corporate charitable donations to charities in the communities in which their plants are located.

(D) The number of automobiles owned per capita in Corland is significantly higher than the number of automobiles owned per capita in Brunia.

(E) Corland imports more automobiles from Brunia than Brunia does from Corland.

9. Bricktown University and Rapids University both have greater numbers of applicants than they have space to admit, and must therefore reject substantial numbers of applicants. The former admits fewer applicants each year than the latter. So we should expect to find that Bricktown University has tougher admissions standards than Rapids University.

Which of the following is an assumption on which the argument depends?

(A) Compared to those who attend Rapids University, a greater percentage of students who attend Bricktown University graduate.

(B) Rapids University does not have significantly greater numbers of applicants than Bricktown University.

(C) Bricktown University's curriculum is more rigorous than Rapids University's.

(D) The percentage of accepted applicants who actually attend Rapids University is less than the percentage who attend Bricktown University.

(E) Rapids University students tend to be better prepared for university study than Bricktown University students.

10. Market researcher: In the next decade, the number of people over 45 will grow and the number of people under 45 will decline. Because last year people over 45 had higher average coffee consumption than people under 45, whereas people under 45 had higher average cola consumption than people over 45, we conclude that, overall, coffee consumption will rise relative to cola consumption over the next decade.

Which of the following, if true, most seriously weakens the market researcher's argument?

(A) Once people acquire a taste for coffee, they are unlikely to significantly reduce their coffee consumption.

(B) A person's choice of beverage typically varies according to the time of day and the activity in which the person is engaged.

(C) Most people develop their beverage drinking habits in their 20's and change them little subsequently.

(D) Overall, coffee consumption is currently somewhat higher than cola consumption.

(E) Both coffee and cola consumption are currently higher among people aged 25 to 45 than among people under 25.

1. Scientists recently tested samples of most major brands of prepared baby food and found that more than half the strained fruit and vegetable products tested

contained pesticide residues. But in all cases the pesticides were at levels well below established government limits, so there is no danger of ill effects from pesticide residues for infants who eat these foods.

Which of the following, if true, most seriously weakens the argument?

(A) All the major brands of prepared baby food had at least one product that was found to contain pesticides.

(B) Homemade baby food made with fruits and vegetables that have been carefully washed contains far lower levels of pesticide residues than do the commercial baby food products that were tested.

(C) The likelihood that a product would be found to contain pesticides did not depend on the type of fruit or vegetable that product contained.

(D) Short-term exposure to pesticide residues in food rarely has long-term effects.

(E) The government limits were established solely on the basis of research about adult tolerances for levels of pesticide residues in food.

2. A breakfast consisting of a bowl of Sweet and Crunchy cereal in milk, as well as a slice of toast and a glass of juice, provides the recommended daily amount of ten essential nutrients. So Sweet and Crunchy cereal is nutritious as well as delicious.

The answer to which of the following questions would be most helpful in assessing the support the evidence provided in the advertisement gives to the conclusion?

(A) Are there any essential nutrients that cannot be obtained from a breakfast consisting of milk, toast, juice, and Sweet and Crunchy cereal?

(B) Does Sweet and Crunchy cereal provide any nutrient for which no recommended daily amount has been established?

(C) How does the price of Sweet and Crunchy cereal compare to that of other cereals that provide the same nutrients that Sweet and Crunchy cereal provides?

(D) What is the recommended daily amount of the ten nutrients provided by a breakfast consisting of Sweet and Crunchy cereal, milk, toast, and juice?

(E) In the breakfast described, does the combination of milk, toast, and juice provide the recommended daily amount of the ten nutrients?

3. There are several known versions of the thirteenth-century book describing a journey Marco Polo of Venice supposedly made to China, yet none contains any description of the Great Wall of China. Since Marco Polo would have had to cross the Great Wall to travel the route described and since the book reports in detail on other, less notable, structures, the omission of the Great Wall strongly suggests that Marco Polo never did actually travel to China.

Which of the following, if true, most strengthens the argument?

(A) The original manuscript of the book describing Marco Polo's supposed journey is no longer in existence.

(B) In his travels, Marco Polo certainly visited his family's trading posts on the Black Sea, where he would have had contact with people who had traveled to the various parts of China that the book describes.

(C) Many notable structures in China that are described in the book were hundreds of years old at the time Marco Polo supposedly traveled to China.

(D) At some places along the Great Wall, a traveler crossing the Wall could do so without realizing that it was an enormous structure.

(E) Certain pieces of accurate information about thirteenth-century China are contained in some of the known versions of the book but not in all known versions.

4. In 1990, the country of ZYX had very poor living conditions in rural areas. Only 45 percent of rural inhabitants had access to clean water, but by 2005, more than 70 percent of rural inhabitants had such access. Clearly, the country has made many effective improvements in its rural infrastructure.

Which of the following statements about ZYX between 1990 and 2005, if true, most weakens the reasoning above?

(A) Living conditions in the cities have improved much more substantially than they have in rural areas.
(B) Living conditions in rural areas have been improved by other infrastructure projects besides projects providing increased access to clean water.
(C) The standards for clean water have not changed substantially since 1990.
(D) Many cities in ZYX have experienced substantial population growth as rural inhabitants move to cities in search of work.
(E) More than 90 percent of the inhabitants in the cities of XYZ have access to clean water.

5. After recent hurricanes produced tremendous damage there, Mardin County revised its building code. In doing so, Mardin imposed stricter standards for roofing but not for windows. The revision will undoubtedly reduce the principal cause of property damage, namely rainwater coming through damaged roofs. However, the revision will do little to prevent personal injuries, since, while many injuries are caused by flying debris breaking through windows, very few are caused by roof collapse or rainwater.

Which of the following, if true, most seriously weakens the argument?

(A) For many years Mardin County's building code has imposed a stricter standard for windows than that imposed by most counties that are frequently struck by hurricanes.

(B) The hurricanes that recently hit Mardin County were not as severe as some hurricanes that have hit there in the past 100 years.

(C) Much of the flying debris carried by the winds of a hurricane is roof material that has become dislodged.

(D) During a hurricane, broken windows in a house cause increased pressure inside the house that can, in turn, cause the roof of the house to become damaged.

(E) Mardin County's building code applies to the repair of houses in addition to the construction of new houses.

6. The more viewers a television show attracts, the greater the advertising revenue the show generates. The television network Vidnet's most popular show, *Starlight*, currently earns the network's highest profit, but next year, because of unavoidable increases in production costs, its profits are projected to fall to below the average for Vidnet shows. Therefore, Vidnet would earn greater profits overall if replaced *Starlight* with a show of average popularity and production costs.

Which of the following, if true, most seriously weakens the argument?

(A) The average profit of Vidnet shows has increased in each of the last three years.

(B) Shows that occupy time slots immediately before and after very popular shows tend to have far more viewers than they otherwise would.

(C) *Starlight* currently has the highest production costs of all Vidnet shows.

(D) Last year Vidnet lost money on a weekly show that was substantially similar to *Starlight* and was broadcast on a different day from *Starlight*.

(E) Even if, as a result of increased production costs, *Starlight* becomes less profitable than the average for Vidnet shows, it will still be more profitable than the average for television shows of all networks combined.

7. Mangroves are salt-tolerant trees that form dense masses along sheltered tropical shorelines, including those on the island of Kalami. On Kalami, traditional uses of mangrove wood are quite limited, and no significant fishing goes on in the mangrove swamps. Therefore, the islanders would reap substantial economic gain by accepting a recent offer made by a mainland woodchip company to pay the islanders a fair price for the right to harvest the wood in their mangrove swamps.

Which of the following, if true, most seriously weakens the argument?

(A) Most of the species of fish that are important to the Kalami islanders' fishing industry depend on mangrove swamps as spawning and nursery areas.

(B) A significant portion of the money that Kalami islanders might receive from the wood chip company would be devoted to modernizing the island's fishing fleet.

(C) There is not enough indigenous forest left in the interior of Kalami to be of interest to the timber industry or even the woodchip industry.

(D) Mangroves do not thrive on the portion of the coastline of Kalami that is oriented toward the open ocean.

(E) Wood whose primary use is for woodchips is considerably less valuable than wood that is used in construction or in furniture making.

8. Online shopping provides consumers with many product alternatives and a vast amount of product information. But purchasing common, everyday items—convenience goods—requires little brand comparison, planning, or new information. Consumers know exactly what to buy, and where to buy it, based on past shopping experience. So, in relation to convenience goods, online shopping has little advantage over traditional shopping.

The argument above is most vulnerable to the criticism that it

(A) fails to take into account that the most significant segment of online shopping sales is not convenience goods

(B) presumes, without providing justification, that convenience goods are typically products of a kind that everybody purchases at some time

(C) presumes, without providing justification, that online shopping has no significant advantages that are relevant besides product choice and product information

(D) fails to consider the advantages of traditional shopping in relation to certain categories of goods other than convenience goods

(E) fails to take into account that even ordinary, everyday items are sometimes not conveniently available to purchase

9. Most proposed legislation is introduced by legislators who, seeking to serve their own personal interests, stand to benefit if the legislation is passed into law. However, since most proposed legislation is not passed into law, most attempts by legislators to serve their own personal interests are unsuccessful.

The reasoning in the argument above is vulnerable to criticism on which of the following grounds?

(A) It does not take into account the fact that legislation can be proposed by people other than legislators.

(B) It does not acknowledge that some proposed legislation is introduced by legislators who are not attempting to serve their own personal interests.

(C) It ignores the possibility that proposed legislation that serves a legislator's personal interests might also serve the interests of other people.

(D) It neglects to indicate what percentage of proposed legislation introduced by legislators is passed into law.

(E) It falls to rule out explicitly the possibility that legislators successfully serve their own personal interests by means other than proposed legislation.

10. In Valgin, gasoline is sold in different grades, the most expensive being premium gasoline. Only about 10 percent of gasoline-powered vehicles sold in Valgin require premium gasoline, and for other vehicles premium provides no advantage. Nevertheless, nearly 15% of gasoline sold there is premium. So presumably a number of Valgian drivers are wasting money when they buy gasoline.

Which of the following, if true, most strengthens the argument?

(A) Buyers of vehicles that require premium gasoline do not drive substantially more miles per year than other drivers.

(B) More of the intermediates grade of gasoline is sold than of premium gasoline.

(C) Most of the vehicles that require premium gasoline are high-performance vehicles and are relatively expensive.

(D) Premium gasoline costs almost 25 percent more than least expensive grade of gasoline.

(E) Some very old vehicles that did not require premium gasoline when they were new now do require it to run smoothly.

第三组

1. Environmentalist: Several thousand chemicals are found in our nation's drinking water, and the health effects of most of them are not completely known. Yet, in the past fifteen years, the government has added only one new chemical to the two dozen or so that it regulates in drinking water. Thus, the government's regulation of chemicals in drinking water is not sufficient to protect public health.

Finding the answer to which of the following questions is most likely to be the most helpful in evaluating the environmentalist's argument?

(A) How many of the chemicals the government currently regulates in drinking water have ever posed a risk to public health when found at high concentrations?

(B) By what means other than regulating chemicals in drinking water does the government work to protect public health?

(C) How many chemicals have been present in significant concentrations in the nation's drinking water for more than fifteen years while not being subject to government regulation?

(D) Why are health effects of most chemicals found in the nation's drinking water not completely known?

(E) Are there two or more chemicals that, at the concentrations present significantly often in the nation's drinking water over the last 15 years, can damage human health?

2. For five years, every auto owner who has switched to Taske Auto Insurance has gotten a rate from 2 to 20 percent less than what they had been paying someone else for the same level of coverage. So clearly, if you are insured by someone else and you call Taske, you can be confident of this: The rate Taske offers you will save you at least 2 percent of what you currently pay!

Which of the following, if true, points to a flaw in the argument in the advertisement?

(A) Current Taske customers with the same level of automobile insurance coverage as each other sometimes pay different rates for that coverage.

(B) Of the people who switched to Taske, the vast majority achieved savings that were closer to 2 percent than to 20 percent.

(C) A large proportion of the auto owners who were quoted auto insurance rates by Taske chose not to switch their insurance to Taske.

(D) Some of Taske's competitors do not offer all of the different levels of coverage offered by Taske.

(E) Taske has been offering automobile insurance for only five years.

3. In Brindon County, virtually all of the fasteners—such as nuts, bolts, and screws—used by workshops and manufacturing firms have for several years been supplied by the Brindon Bolt Barn, a specialist wholesaler. In recent months many of Brindon County's workshops and manufacturing firms have closed down, and no new ones have opened. Therefore, the Brindon Bolt Barn will undoubtedly show a sharp decline in sales volume and revenue for this year as compared to last year.

The argument depends on assuming which of the following?

(A) Last year the Brindon Bolt Barn's sales volume and revenue were significantly higher than they had been the previous year.

(B) The workshops and manufacturing firms that have remained open have a smaller volume of work to do this year than they did last year.

(C) Soon the Brindon Bolt Barn will no longer be the only significant supplier of fasteners to Brindon County's workshops.

(D) The Brindon Bolt Barn's operating expenses have not increased this year.

(E) The Brindon Bolt Barn is not a company that gets the great majority of its business from customers outside Brindon County.

4. In the United States, many companies provide health care insurance for their employees. More and more of those companies are implementing programs promoting healthy lifestyles, programs that focus on factors associated with rising medical costs, such as high blood pressure. When successful, such programs help reduce the costs of health care, and thus of health insurance. In Canada, however, the government pays for health care. Clearly, therefore, Canadians companies have no financial interest to implement similar programs.

Which of the following, if true, most seriously weakens the argument?

(A) When employees miss work because of illness, their employer typically incurs expenses related to finding temporary substitutes for those employees.

(B) Many United States companies have found that certain programs promoting healthy lifestyles improve employees' health far more than other such programs do.

(C) Some companies introduce programs promoting healthy lifestyles without first analyzing whether, in this specific case, there will be a net cost savings.

(D) For people who seek employment with a company, the presence or absence of a program promoting healthy lifestyles is not generally significant consideration.

(E) The cost of specific medical treatments is generally lower in Canada than it is in the United States.

5. Solar radiation is a cause of skin cancer. In the United States, skin cancer is more common on the left side of the face than on the right. Some dermatologists hypothesize that this difference is due to drivers in the United States being directly exposed to solar radiation on the left side of the face more often than on the right, since cars in the United States have the driver's side window to the driver's left.

Which of the following, if true, most strongly supports the dermatologists' hypothesis?

(A) Many people in the United States who develop skin cancer do not regularly drive a car.

(B) When the windows of a car are closed, they absorb much of the harmful solar radiation to which drivers would otherwise be exposed.

(C) In countries where cars have the driver's side window to the driver's right, skin cancer is more common on the right side of the face.

(D) In the United States, most drivers who are diagnosed with skin cancer continue to drive after being diagnosed.

(E) Many people who spend more time driving than the average also spend more time out of doors than the average.

6. Science reporter: In an experiment, scientists fed one group of mice an ordinary diet and fed another group a diet fortified with the microbes found in yogurt. The mice then explored a maze made up of tunnels and uncovered paths. The mice on the fortified diet ventured out of the tunnels twice as often as the other mice did, suggesting that the mice on the fortified diet felt less anxiety about being out in the open. This shows that the differences in the mice's diets affected their anxiety levels.

Which of the following, if true, would most strongly support the science reporter's conclusion?

(A) In addition to having the added microbes, the fortified diet differed from the ordinary diet by having significantly fewer calories.

(B) The microbes found in yogurt produce a chemical known to lower anxiety levels.

(C) In a second experiment conducted to check the results of the first, the diets of the two groups of mice were switched, and then the mice ran the maze again.

(D) Dietary changes can affect mice's sensitivity to light and thereby affect the strength of their preference for dark, enclosed spaces.

(E) The anxiety levels of mice affect their behavior in many ways that the scientists could have easily observed during the experiment.

7. Athletic coach: A recent survey of secondary school athletes revealed that those who use vitamins, protein powder, and other similar nutritional supplements

were much more likely to be successful in athletics than those who do not. This is clear evidence that such supplements are indeed effective.

The athletic coach's argument is most vulnerable to criticism on which of the following grounds?

(A) It overlooks the possibility that athletes might not reveal anything that might be perceived as giving them an unfair advantage.

(B) It overlooks the possibility that athletes who are more likely to achieve athletic success even without nutritional supplements are more likely to use such supplements.

(C) It takes for granted that athletes engaged in team sports would be affected by supplements in the same way as competitors in nonteam sports.

(D) It fails to adequately address the possibility that among those who achieve athletic success, some might be more fit than others regardless of whether they take supplements or not.

(E) It fails to adequately address the possibility that what is true of athletes as a group might not be true of certain individual athletes.

8. Vitamin D is obtained directly from food and synthesized by means of exposure to sunlight. Among people who eat similar diets and live in areas with equal numbers of sunny days, those who are in poor health generally have lower levels of vitamin D in their blood than do those in good health.

Hypothesis: Poor health can affect the body's ability to utilize available vitamin D.

Which of the following, if true, most seriously undermines the hypothesis?

(A) People who are in poor health tend to have their blood tested more often than do people in good health.

(B) Some people in good health also have low levels of vitamin D in their blood.

(C) People who are in poor health tend to stay indoors more frequently.

(D) Experts disagree on the minimum level of vitamin D required to maintain a healthy body.

(E) People in poor health who eat a diet that is relatively low in vitamin D also generally have relatively low levels of vitamin D in their blood.

9. Fearing competition from stores that rent video games for home use, owners of stores that sell video games lobbied for protective legislation. Citing as a precedent legislation that postpones home film rentals until one year after general release to theaters, the video sellers proposed as an equitable solution a plan that would postpone rental of any video game until it had been available for sale for one year.

Which of the following, if true, would support an objection by owners of video rental stores that the fairness of the proposed legislation is not supported by the precedent cited?

(A) Although the daily rental fee for home use of films is generally the same as that for video games, the average rental period for a video game is longer than that for a film.

(B) Film rentals for home use tend to be particularly strong during the first year after a film has been made available for both sale and rental at video rental stores.

(C) Most people are interested in playing only the latest video games and tend to find video games that have been available for over a year unappealing whether they have played them before or not, whereas films maintain their appeal far longer after their release.

(D) People who rent video games tend to play them by themselves, whereas people who rent films invite friends and neighbors to watch.

(E) A slight decline in revenues from films that have been recently released in theaters has been attributed to the growing market for rentals of films for home use.

10. Newspaper article: Currently there are several prominent candidates for the position of governor of Province X. The prime minister has endorsed one of these candidates, Ann Shuh. The last candidate the prime minister endorsed for a provincial governorship won the position over strong rival candidates. Therefore, Ann Shuh will probably win the governorship of Province X.

The newspaper article's argument is most vulnerable to criticism on which of the following grounds?

(A) It asserts a sweeping generalization about a group on the basis of evidence regarding only a single, possibly unrepresentative, member of that group.

(B) It takes for granted that if Ann Shuh wins the governorship, the prime minister's endorsement will have been the main cause of her victory.

(C) It relies on a single point of similarity between two situations, without adequately addressing the possibility of relevant differences between those situations.

(D) It confuses a claim about a past event with a prediction about a similar, future event.

(E) It overlooks the possibility that the last candidate the prime minister endorsed for a provincial governorship could not have won the position without that endorsement.

答案及解析

1. The insecticide occurs throughout the plant, including its pollen. Maize pollen is dispersed by the wind and frequently blows onto milkweed plants that grow near maize fields. Caterpillars of monarch butterflies feed exclusively on milkweed leaves. When these caterpillars are fed milkweed leaves dusted with pollen from modified maize plants, they die. Therefore, by using genetically modified maize, farmers put monarch butterflies at risk.

Which of the following would it be most useful to determine in order to evaluate the argument?

(A) Whether the natural insecticide is as effective against maize-eating insects as commercial insecticides typically used on maize are

(B) Whether the pollen of genetically modified maize contains as much insecticide as other parts of these plants

(C) Whether monarch butterfly caterpillars are actively feeding during the part of the growing season when maize is releasing pollen

(D) Whether insects that feed on genetically modified maize plants are likely to be killed by insecticide from the plant's pollen

(E) Whether any maize-eating insects compete with monarch caterpillars for the leaves of milkweed plants growing near maize fields

类别：普通预测推理

前提：通过使用转基因玉米。

结论：农民将帝王蝶置于危险之中。

推理：结论"农民将帝王蝶置于危险之中"的充分条件应为"帝王蝶会吃这些转基因的玉米"。

选项分析

（A）对吃玉米的昆虫来说，天然杀虫剂与通常用于玉米的商业杀虫剂是否一样有效。对比杀虫剂的效果与讨论无关。

（B）转基因玉米的花粉是否和这些植物的其他部分一样含有大量的杀虫剂。对比杀虫剂的量与讨论无关。

（C）正确。在玉米释放花粉的生长季节，帝王蝶毛毛虫是否在积极觅食。本选项讨论了“使用转基因玉米”能否＝毛毛虫会吃到。如果在花粉释放季节，毛毛虫不怎么觅食，那蝴蝶就不会受到威胁。

（D）以转基因玉米植物为食的昆虫是否有可能被来自该植物花粉的杀虫剂杀死。与吃转基因玉米的昆虫无关。

（E）是否有吃玉米的昆虫与帝王蝶毛毛虫争夺生长在玉米地附近的乳草植物的叶子。与是否有竞争者无关。无论有无竞争者，都不影响毛毛虫自己吃到有毒的花粉。

2. Banking-industry analysts have been predicting the end of the paper check as a method of making payments, now that electronic transfer of funds from one bank account to another is becoming increasingly common. To process paper checks, banks rely on companies that charge a significant amount for each check they process. Therefore, banks would be well advised to encourage the replacement of paper checks with electronic transfers as the standard method of payment.

Which of the following, if true, most seriously weakens the argument?

(A) Even consumers who currently make use of electronic transfers to pay their regularly recurring bills still commonly use paper checks to make payments for occasional purchases.

(B) Some consumer checking accounts pay the account holder interest on checking account balances.

(C) Government agencies that make regular payments to individuals are increasingly requiring that the people to whom they make such payments receive the funds by electronic transfers into a bank account.

(D) Banks are able to charge their customers substantial fees for many of the services associated with paper-check transactions.

(E) The number of paper-check transactions has decreased only somewhat even though electronic fund transfers have become more common.

类别：普通预测推理

前提：银行要为纸质支票付大量费用。

结论：银行会支持用电子转账取代纸质支票作为标准的支付方法。

推理：结论“银行会支持用电子转账取代纸质支票作为标准的支付方法”的充分条件应为“电子转账无须付大量费用且银行没有其他理由依赖纸质支票”。

选项分析：

(A) 即使是目前使用电子转账来支付经常性账单的消费者，仍然经常使用纸质支票来支付偶尔的消费。偶尔的纸质支票消费与讨论无关。

(B) 一些消费者支票账户银行向账户持有人支付支票账户余额的利息。这是支票给消费者带来的好处。

(C) 定期向个人付款的政府机构越来越多地要求他们的付款对象以电子转账的方式将资金存入他们的银行账户。没有提到纸质支票的好处或者电子支票的缺点。

(D) 正确。银行能够就许多与纸质支票交易相关的服务向客户收取大量费用。本选项给出了银行可能会继续依赖纸质支票的理由。

(E) 尽管电子资金转账越来越普遍，但纸质支票交易的数量只略有减少。没有提到纸质支票的好处或者电子支票的缺点。

3. Because of steep increases in the average price per box of cereal over the last 10 years, overall sales of cereal have recently begun to drop. In an attempt to improve sales, one major cereal manufacturer reduced the wholesale prices of its cereals by 20 percent. Since most other cereal manufacturers have announced that they will follow suit, it is likely that the level of overall sales of cereal will rise significantly.

Which of the following would it be most useful to establish in evaluating the argument?

(A) Whether the high marketing expenses of the highly competitive cereal market led to the increase in cereal prices

(B) Whether cereal manufacturers use marketing techniques that encourage brand loyalty among consumers

(C) Whether the variety of cereals available on the market has significantly increased over the last 10 years

(D) Whether the prices that supermarkets charge for these cereals will reflect the lower prices the supermarkets will be paying the manufacturers

(E) Whether the sales of certain types of cereal have declined disproportionately over the last 10 years

类别：普通预测推理

前提：谷物批发价格减少 20%。

结论：谷物销量增加。

推理：结论“谷物销量增加”的充分条件应为“谷物的售卖价格下降且需求不变（甚至增加）”。批发价格并不一定等于真实售价。

选项分析：

(A) 竞争激烈的谷物市场高昂的营销费用是否会导致谷物的价格上涨。虽然本选项涉及了“价格”这个词项，但是谷物市场的竞争是一直存在的，所以如果批发价下调了，整体市场的价格还是会下调的。

(B) 谷物制造商是不是用了一些营销技术来鼓励在顾客中建立品牌忠诚度。无论谷物制造商有没有类似的营销技术，其都不能反驳销量将要上升的这个事实。

(C) 谷物的种类在过去十年中是不是显著增加了。谷物种类和价格无关。

(D) 正确。超市所售卖的谷物价格是否能反映超市向谷物制造商支付的价格低。显然，本选项直接讨论了谷物的售卖价格。

(E) 过去十年中，某些种类的谷物的销售量是否出现了不成比例的下降。某种谷物的情况无法说明所有谷物的平均情况。

4. Wind farms, which generate electricity using arrays of thousands of wind-powered turbines, require vast expanses of open land. County X and County Y have similar terrain, but the population density of County X is significantly higher than that of County Y. Therefore, a wind farm proposed for one of the two counties should be built in County Y rather than in County X.

Which of the following, if true, most seriously weakens the planner's argument?

(A) County X and County Y are adjacent to each other, and both are located in the windiest area of the state.

(B) The total population of County Y is substantially greater than that of County X.

(C) Some of the electricity generated by wind farms in County Y would be purchased by users outside the county.

(D) Wind farms require more land per unit of electricity generated than does any other type of electrical-generation facility.

(E) Nearly all of County X's population is concentrated in a small part of the county, while County Y's population is spread evenly throughout the county.

类别：类比推理

前提：X 县的人口密度高于 Y 县的人口密度。

结论：应该把风力发电厂建造在Y县而不是Y县。

推理：考虑X县和Y县其他相关相似点的缺失。

选项分析：

(A) X县和Y县彼此毗邻，两者都是所属州风力最大的地区。本选项和空旷地方有多少没有联系。

(B) Y县的人口总量要远远大于X县。人口多少和地方是否空旷没有关系（还要考虑地理面积）。

(C) Y县风力发电厂发出的部分电能将被该县以外的用户买走。本选项提及的是电能最后的归属问题，和地方是否空旷没有关系。

(D) 产生单位电能的情况下，风力发电厂比其他种类的发电设施需要的土地资源更多。本选项提及的是风力发电厂和其他发电设施相比需要土地的情况，和三个维度无关。

(E) 正确。几乎所有X县的人都聚集在该县很小的一部分地方生活，而在Y县，大家都是分散居住的。本选项直接指出了两个县的不同。虽然X县的人口密度大，但是若X县的人大都住在一起，那么空旷的地方可能反而要多于Y县。

5. Because it was long thought that few people would watch lengthy televised political messages, most televised political advertisements, like commercial advertisements, took the form of short messages. Last year, however, one candidate produced a half-hour-long advertisement. At the beginning of the half-hour slot a substantial portion of the viewing public had tuned into that station. Clearly, then, many more people are interested in lengthy televised political messages than was previously thought.

Which of the following, if true, most seriously weakens the argument?

(A) The candidate who produced the half-hour-long advertisement did not win election at the polls.

(B) The half-hour-long advertisement was widely publicized before it was broadcast.

(C) The half-hour-long advertisement was aired during a time slot normally taken by one of the most popular prime-time shows.

(D) Most short political advertisements are aired during a wide range of programs in order to reach a broad spectrum of viewers.

(E) In general a regular-length television program that features debate about current political issues depends for its appeal on the personal qualities of the program's moderator.

类别：归因推理

前提：在节目开始的时候很多人调进该电视频道。

结论：更多的人对长的政治信息感兴趣。

推理："更多的人对长的政治信息感兴趣"的其他证据不存在或给出其他能解释"很多人会调台"的原因。

选项分析：

(A) 那个播放了半小时长的政治广告的候选者没有赢得选举。候选者是否赢得选举与观众是否会调台专门去看他无关。

(B) 在这条广告播放以前，这条半小时长的广告已经众所周知了。观众是否知道电视台要播放广告和他们是否喜欢这条广告没有关系。

(C) 正确。这条半小时长的广告是在一个原本播放最受欢迎的表演的黄金时段播放的。观众可能是误以为还是会播放受欢迎的表演所以调到这个电视频道，而不是因为喜欢这条广告。

(D) 大部分短的政治广告都插播在许多种节目中，这样做的目的是吸引尽量多的观众。本选项描述的是短政治广告的情况，而原文是关于长政治广告的，可以排除。

(E) 一般情况下，讨论当今政治话题的正常长度的电视节目的特色取决于该节目对主持人个人品质的要求。本选项讨论的是政治节目质量的问题，与前提和结论都无关。

6. Two computer companies, Garnet and Renco, each pay Salcor to provide health insurance for their employees. Because early treatment of high cholesterol can prevent strokes that would otherwise occur several years later, Salcor encourages Garnet employees to have their cholesterol levels tested and to obtain early treatment for high cholesterol. Renco employees generally remain with Renco only for a few years, however. Therefore, Salcor lacks any financial incentive to provide similar encouragement to Renco employees.

Which of the following, if true, most seriously weakens the argument?

(A) Early treatment of high cholesterol does not eliminate the possibility of a stroke later in life.

(B) People often obtain early treatment for high cholesterol on their own.

(C) Garnet hires a significant number of former employees of Renco.

(D) Renco and Garnet have approximately the same number of employees.

(E) Renco employees are not, on average, significantly younger than Garnet employees.

类别：普通预测推理

前提：Renco 员工只在 Renco 待几年。

结论：Salcor 缺少给 Renco 员工提供相似鼓励的财务动机。

推理：结论“Salcor 缺少财务动机”的充分条件应为“Salcor 永远不会负责 Renco 员工的医疗支出”。但前提只讲了 Renco 员工只在 Renco 待几年。

选项分析：

(A) 早期治疗高胆固醇并不能消除日后中风的可能。本选项描述的是治疗后的效果，和讨论无关。

(B) 人们经常自己去对高胆固醇进行早期治疗。本选项描述的是员工自己是否会治疗，和讨论无关。

(C) 正确。Garnet 雇用很多 Renco 的前员工。由于 Garnet 和 Renco 都是由 Salcor 提供保险的，所以如果 Garnet 雇用很多 Renco 的前员工，这些曾经是 Renco 的员工虽然可能在 Renco 不会发病，但是几年后到了 Garnet 有可能会发病，Garnet 依然处在 Salcor 的保额范围内，所以显然，Salcor 还是会让 Renco 的员工去检测他们的胆固醇。

(D) Renco 和 Garnet 的员工数量相同。无论员工数量是否相同，都不会影响到 Salcor 提供的保险的问题。

(E) 平均而言，Renco 的员工不会比 Garnet 的员工年轻。无论员工是否年轻，只要胆固醇高，其后几年就有中风的可能。

7. In the United States, of the people who moved from one state to another when they retired, the percentage who retired to Florida has decreased by three percentage points over the past ten years. Since many local businesses in Florida cater to retirees, this decline is likely to have a noticeably negative economic effect on these businesses.

Which of the following, if true, most seriously weakens the argument?

(A) Florida attracts more people who move from one state to another when they retire than does any other state.

(B) The number of people who move out of Florida to accept employment in other states has increased over the past ten years.

(C) There are far more local businesses in Florida that cater to tourists than there are local businesses that cater to retirees.

（D）The total number of people who retired and moved to another state for their retirement has increased significantly over the past ten years.

（E）The number of people who left Florida when they retired to live in another state was greater last year than it was ten years ago.

类别：普通预测推理

前提：退休后移居的人中，选择搬到佛罗里达州的人减少了3%。

结论：佛罗里达州迎合退休人员需求的企业会受到负面影响。

推理：结论“企业会受到负面影响”的充分条件应为“选择搬到佛罗里达州的人数有所减少”。但前提提到的是百分比减少。很明显，百分比和人数不是一回事，我们需要考虑总人数（基数）的情况。

选项分析：

（A）佛罗里达州吸引的退休后从一个州搬到另一个州的人比其他任何一个州都多。和其他州做比较没有意义，不管比其他州多还是少，都和佛罗里达州的企业会不会受影响无关。

（B）在过去的十年里，从佛罗里达州搬到其他州工作的人数量有所增加。讨论的是退休后“搬去”佛罗里达州的人，和从佛罗里达州“搬出来”的人无关。

（C）在佛罗里达州，为游客服务的当地企业远远多于为退休人员服务的当地企业。讨论的是“迎合退休人员需求的企业”，和旅游业无关。

（D）正确。退休后搬到另一个州养老的总人数在过去十年里显著增加。此选项给出了基数情况。

（E）退休后从佛罗里达州搬到另一个州养老的总人数在过去十年里显著增加。讨论的是退休后“搬去”佛罗里达州的人，和从佛罗里达州“搬出来”的人无关。

8. Although automobile manufacturers in Brunia charge the same prices for automobiles as do automobile manufacturers in the neighboring country of

Corland and the tax structures of the two countries are nearly identical, the hourly wage that Brunian automobile manufacturers pay their employees is much lower. Therefore, unless automobile manufacturers in Brunia pay substantially more for their raw materials, they make higher profits than do automobile manufacturers in Corland.

Which of the following, if true, most seriously weakens the argument?

(A) Automobile manufacturers in Corland pay more for some raw materials than do automobile manufacturers in Brunia.

(B) In Brunia but not in Corland, automobile manufacturers are required by law to pay their employees health care costs in addition to their wages.

(C) In Corland but not in Brunia, most automobile manufacturers restrict their corporate charitable donations to charities in the communities in which their plants are located.

(D) The number of automobiles owned per capita in Corland is significantly higher than the number of automobiles owned per capita in Brunia.

(E) Corland imports more automobiles from Brunia than Brunia does from Corland.

类别：普通预测推理

前提：布鲁尼亚的汽车制造商与邻国科兰的汽车制造商销售的汽车价格相同，而且两国的税收结构也几乎相同，但布鲁尼亚汽车制造商支付给员工的小时工资却低得多。

结论：除非布鲁尼亚的汽车制造商为其原材料支付更多的费用，否则他们的利润会高于科兰的汽车制造商。

推理：结论讨论的是“利润”，前提提到的是收入（汽车价格）和部分成本（税收、人工成本等）。但决定利润的充分条件应是全部收入减去全部成本。所以，答案需要指出存在其他收入或成本。

选项分析：

(A) 科兰的汽车制造商为某些原材料支付的费用比布鲁尼亚的汽车制造商高。原材料不在讨论范围内。

(B) 正确。在布鲁尼亚而不是在科兰，法律要求汽车制造商除了支付员工工资外，还要支付员工的医疗费用。本选项提供了一个额外的成本。

(C) 在科兰，而不是在布鲁尼亚，大多数汽车制造商将他们公司的慈善捐款只捐给其工厂所在社区的慈善机构。慈善捐款的去向与讨论无关。

(D) 科兰的人均汽车拥有量明显高于布鲁尼亚的人均汽车拥有量。汽车拥有量与讨论无关。

(E) 科兰从布鲁尼亚进口的汽车比布鲁尼亚从科兰进口的汽车多。汽车进口量与讨论无关。

9. Bricktown University and Rapids University both have greater numbers of applicants than they have space to admit, and must therefore reject substantial numbers of applicants. The former admits fewer applicants each year than the latter. So we should expect to find that Bricktown University has tougher admissions standards than Rapids University.

Which of the following is an assumption on which the argument depends?

(A) Compared to those who attend Rapids University, a greater percentage of students who attend Bricktown University graduate.

(B) Rapids University does not have significantly greater numbers of applicants than Bricktown University.

(C) Bricktown University's curriculum is more rigorous than Rapids University's.

(D) The percentage of accepted applicants who actually attend Rapids University is less than the percentage who attend Bricktown University.

(E) Rapids University students tend to be better prepared for university study than Bricktown University students.

类别：普通预测推理

前提：前者（Bricktown 大学）每年录取的申请人比后者（Rapids 大学）少。

结论：Bricktown 大学的录取标准比 Rapids 大学更严格。

推理：结论“录取标准更严格”的充分条件应为“录取率更低”。但前提讲的是被录取人数。所以，答案可以指出申请人数的情况。

选项分析：

(A) 与就读于 Rapids 大学的学生相比，就读于 Bricktown 大学的学生毕业的比例更高。毕业率与录取标准无关。

(B) 正确。Rapids 大学的申请者人数并不比 Bricktown 大学多很多。

(C) Bricktown 大学的课程比 Rapids 大学的课程更严格。课程严格程度与录取标准无关。

(D) 实际就读于 Rapids 大学的被录取申请者的比例低于就读于 Bricktown 大学的比例。最终实际有多少人接到 offer 与讨论无关。

(E) Rapids 大学的学生往往比 Bricktown 大学的学生为大学学习做了更好的准备。与学习准备无关。

10. Market researcher: In the next decade, the number of people over 45 will grow and the number of people under 45 will decline. Because last year people over 45 had higher average coffee consumption than people under 45, whereas people under 45 had higher average cola consumption than people over 45, we conclude that, overall, coffee consumption will rise relative to cola consumption over the next decade.

Which of the following, if true, most seriously weakens the market researcher's argument?

(A) Once people acquire a taste for coffee, they are unlikely to significantly reduce their coffee consumption.

(B) A person's choice of beverage typically varies according to the time of day and the activity in which the person is engaged.

(C) Most people develop their beverage drinking habits in their 20's and change them little subsequently.

(D) Overall, coffee consumption is currently somewhat higher than cola consumption.

(E) Both coffee and cola consumption are currently higher among people aged 25 to 45 than among people under 25.

类别：普通预测推理

前提：去年45岁以上的人的咖啡平均消费量高于45岁以下的人，而45岁以下的人的可乐平均消费量高于45岁以上的人；未来十年，45岁以上的人的数量会增加。

结论：未来十年，咖啡消费量相对于可乐消费量将上升。

推理：结论"咖啡消费量相对于可乐消费量将上升"的充分条件应为"未来十年，45岁以上的人的咖啡平均消费量更高"。但前提讲的是现在的情况，所以，答案可以讲出十年后的情况。

选项分析：

(A) 一旦人们习惯了咖啡的口味，他们就不太可能大幅减少咖啡的消费量。没有指出"年龄"和"饮品消费量"的关系。

(B) 一个人对饮料的选择通常会根据一天中的时间和他所从事的活动而变化。讨论的是"年龄"和"饮品消费量"的关系，与"从事活动"和"饮品消费量"的关系无关。

(C) 正确。大多数人在20多岁时养成了喝饮料的习惯，随后改变不大。如果大家不会改变喝饮料的习惯，那么这些45岁以下的人即便到了十年之后，也依然会喝可乐而不是咖啡。

（D）总的来说，目前咖啡消费量比可乐消费量要高一些。目前的情况与未来十年无关。

（E）目前 25 至 45 岁的人的咖啡和可乐消费量都比 25 岁以下的人高。目前的情况与未来十年无关。

第二组

1. Scientists recently tested samples of most major brands of prepared baby food and found that more than half the strained fruit and vegetable products tested contained pesticide residues. But in all cases the pesticides were at levels well below established government limits, so there is no danger of ill effects from pesticide residues for infants who eat these foods.

Which of the following, if true, most seriously weakens the argument?

(A) All the major brands of prepared baby food had at least one product that was found to contain pesticides.

(B) Homemade baby food made with fruits and vegetables that have been carefully washed contains far lower levels of pesticide residues than do the commercial baby food products that were tested.

(C) The likelihood that a product would be found to contain pesticides did not depend on the type of fruit or vegetable that product contained.

(D) Short-term exposure to pesticide residues in food rarely has long-term effects.

(E) The government limits were established solely on the basis of research about adult tolerances for levels of pesticide residues in food.

类别：普通预测推理

前提：在所有情况下，这些农药的含量都远远低于政府规定的限度。

结论：食用这些食品的婴儿没有受到农药残留的危害。

推理：结论“农药残留对婴儿没有危害”的充分条件为“农药含量足够低以至于不会伤害婴儿”，但原文前提给出的是“政府规定的限度”，所以我们需要指出政府规定的限度不一定足够低。

选项分析：

(A) 所有主要品牌的婴儿食品中至少有一种产品被发现含有杀虫剂。“含有杀虫剂”与“杀虫剂是否会对婴儿造成危害”是两回事。

(B) 用经过仔细清洗的水果和蔬菜制成的自制婴儿食品，其农药残留水平远远低于被测试的商业婴儿食品。此对比与讨论无关。

(C) 一种产品被发现含有杀虫剂的可能性并不取决于该产品所含水果或蔬菜的种类。与讨论无关。

(D) 短期接触食品中的农药残留物很少会产生长期影响。与产生长期还是短期影响无关。

(E) 正确。政府制定的限值完全是基于成人对食品中农药残留水平耐受性的研究。显然，如果政府的标准是为成人设立的，那么婴儿吃含有同等剂量的农药残留物的食品就很可能有害。

2. A breakfast consisting of a bowl of Sweet and Crunchy cereal in milk, as well as a slice of toast and a glass of juice, provides the recommended daily amount of ten essential nutrients. So Sweet and Crunchy cereal is nutritious as well as delicious.

The answer to which of the following questions would be most helpful in assessing the support the evidence provided in the advertisement gives to the conclusion?

(A) Are there any essential nutrients that cannot be obtained from a breakfast consisting of milk, toast, juice, and Sweet and Crunchy cereal?

(B) Does Sweet and Crunchy cereal provide any nutrient for which no recommended daily amount has been established?

(C) How does the price of Sweet and Crunchy cereal compare to that of other cereals that provide the same nutrients that Sweet and Crunchy cereal provides?

(D) What is the recommended daily amount of the ten nutrients provided by a breakfast consisting of Sweet and Crunchy cereal, milk, toast, and juice?

(E) In the breakfast described, does the combination of milk, toast, and juice provide the recommended daily amount of the ten nutrients?

类别：普通预测推理

前提：由一碗加牛奶的甜脆麦片、一片吐司和一杯果汁组成的早餐，可以提供每日推荐的十种基本营养物质。

结论：甜脆麦片营养丰富。

推理：结论“甜脆麦片营养丰富”的充分条件是甜脆麦片可以提供每日推荐的十种基本营养物质，但现在的前提讲的是混合了牛奶、吐司和果汁的甜脆麦片可以提供十种营养物质，所以答案选项应考虑是否是牛奶、吐司和果汁提供的十种营养物质。

选项分析：

(A) 是否有任何必需的营养物质不能从由牛奶、吐司、果汁和甜脆麦片组成的早餐中获得？与不能从组合早餐中获得的物质无关。

(B) 甜脆麦片是否提供了任何没有确定每日推荐量的营养物质？与其他营养物质无关。

(C) 与其他提供和甜脆麦片相同营养物质的麦片相比，甜脆麦片的价格如何？与价格无关。

(D) 由甜脆麦片、牛奶、吐司和果汁组成的早餐所提供的十种营养物质的每日推荐量是多少？与具体的推荐量无关。

(E) 正确。在所述的早餐中，牛奶、吐司和果汁的组合是否提供了十种营养物质的每日推荐量？

3. There are several known versions of the thirteenth-century book describing a journey Marco Polo of Venice supposedly made to China, yet none contains any description of the Great Wall of China. Since Marco Polo would have had to cross the Great Wall to travel the route described and since the book reports in detail on other, less notable, structures, the omission of the Great Wall strongly suggests that Marco Polo never did actually travel to China.

Which of the following, if true, most strengthens the argument?

(A) The original manuscript of the book describing Marco Polo's supposed journey is no longer in existence.

(B) In his travels, Marco Polo certainly visited his family's trading posts on the Black Sea, where he would have had contact with people who had traveled to the various parts of China that the book describes.

(C) Many notable structures in China that are described in the book were hundreds of years old at the time Marco Polo supposedly traveled to China.

(D) At some places along the Great Wall, a traveler crossing the Wall could do so without realizing that it was an enormous structure.

(E) Certain pieces of accurate information about thirteenth-century China are contained in some of the known versions of the book but not in all known versions.

类别：归因推理

前提：马可·波罗必须穿越长城才能走完所描述的路线，而且书中还详细报道了其他不那么引人注目的建筑，以及书中没有对长城进行描写。

结论：马可·波罗实际上从未到过中国。

推理：答案选项可以给出“马可·波罗实际上从未到过中国”的其他证据或说明其他能解释前提的原因不存在。

选项分析：

(A) 描述马可·波罗所谓的旅行的书的原稿已经不存在了。原稿是否存在都不能证明马可·波罗是否真的去过中国。

(B) 正确。在他的旅行中，马可·波罗肯定访问了他家在黑海的贸易站，在那里他可能会与书中描述的那些曾到过中国各地的人有过接触。马可·波罗从未到过中国的另外一个证据可以是“他和一些了解中国的人沟通过才能写出那些文字”。

(C) 书中所描述的许多中国著名建筑，在马可·波罗据说前往中国的时候，已经有几百年的历史了。与建筑有多久的历史无关。

(D) 在长城沿线的一些地方，穿越长城的旅行者可以在没有意识到它是一座巨大建筑的情况下进行穿越。本选项从“根据前提还能推理出什么”给出了一个他因，即可能是因为忽视才没有在书中提到长城，而不是因为没有到访过中国。但本选项是削弱，而题目问的是加强。

(E) 关于 13 世纪中国的某些准确信息包含在该书的一些已知版本中，但不是所有已知版本中。与讨论无关。

4. In 1990, the country of ZYX had very poor living conditions in rural areas. Only 45 percent of rural inhabitants had access to clean water, but by 2005, more than 70 percent of rural inhabitants had such access. Clearly, the country has made many effective improvements in its rural infrastructure.

Which of the following statements about ZYX between 1990 and 2005, if true, most weakens the reasoning above?

(A) Living conditions in the cities have improved much more substantially than they have in rural areas.

(B) Living conditions in rural areas have been improved by other infrastructure projects besides projects providing increased access to clean water.

(C) The standards for clean water have not changed substantially since 1990.

(D) Many cities in ZYX have experienced substantial population growth as rural inhabitants move to cities in search of work.

(E) More than 90 percent of the inhabitants in the cities of XYZ have access to clean water.

类别：归因推理

前提：到 2005 年，超过 70%（过去为 45%）的农村居民能够获得清洁水。

结论：这个国家在农村基础设施方面已经进行了许多有效的改善。

推理：答案选项可以给出“这个国家在农村基础设施方面已经进行了许多有效的改善”的其他证据不存在或给出其他能解释前提的原因。

选项分析：

(A) 城市的生活条件比农村地区有更大的改善。与城市的生活条件无关。

(B) 除了提供更多清洁水的项目外，其他基础设施项目也改善了农村地区的生活条件。本选项给出了另一些结论成立的证据。但本题问的是削弱，本选项是加强，所以不正确。

(C) 自 1990 年以来，清洁水的标准没有实质性的改变。如果标准有改变，则可能是因为标准改变而非基础设施变好而导致能获得清洁水的人口比例更高。但同样，本选项是加强，它排除了一个可能的他因，所以不正确。

(D) 正确。由于农村居民进城寻找工作，ZYX 的许多城市都经历了人口的大幅增长。本选项指出了获得干净水源的农民占比增加的另一个原因——农民数量减少，所以干净水源的数量可能没有变化，只是农民数量变少罢了。

(E) 在 XYZ 的城市中，90% 以上的居民都能获得清洁水。与城市的情况无关。

5. After recent hurricanes produced tremendous damage there, Mardin County revised its building code. In doing so, Mardin imposed stricter standards for roofing but not for windows. The revision will undoubtedly reduce the principal cause of property damage, namely rainwater coming through damaged roofs.

However, the revision will do little to prevent personal injuries, since, while many injuries are caused by flying debris breaking through windows, very few are caused by roof collapse or rainwater.

Which of the following, if true, most seriously weakens the argument?

(A) For many years Mardin County's building code has imposed a stricter standard for windows than that imposed by most counties that are frequently struck by hurricanes.

(B) The hurricanes that recently hit Mardin County were not as severe as some hurricanes that have hit there in the past 100 years.

(C) Much of the flying debris carried by the winds of a hurricane is roof material that has become dislodged.

(D) During a hurricane, broken windows in a house cause increased pressure inside the house that can, in turn, cause the roof of the house to become damaged.

(E) Mardin County's building code applies to the repair of houses in addition to the construction of new houses.

类别：普通预测推理

前提：许多人受伤是由飞溅的碎片击穿窗户造成的，而很少是由屋顶坍塌或雨水造成的。

结论：该修订（对屋顶实施更严格的标准）对防止人受伤几乎没有作用。

推理：结论“对防止人受伤几乎没有作用”的充分条件为“屋顶不会以任何形式给人带来伤害”，但原文前提只给出了坍塌这一种情况，所以我们需要指出屋顶可以在其他情况下对人造成伤害。

选项分析：

(A) 多年来，马尔丁县的建筑规范对窗户的要求比大多数经常遭受飓风袭击的县的标准更严格。与窗户的相关规范无关。

（B）最近袭击马尔丁县的飓风并不像过去一百年里袭击那里的一些飓风那样强烈。与之前的飓风做比较没有意义。

（C）正确。飓风携带的大部分飞散的碎片是已经脱落的屋顶材料。本选项说明了屋顶能造成其他伤害。

（D）在飓风期间，房屋的窗户破裂，导致房屋内部压力增加，这反过来又会导致屋顶受损。此选项指出破坏先从窗户开始，与结论无关。

（E）马尔丁县的建筑规范除了适用于建造新房外，还适用于房屋的维修。与维修房屋无关。

6. The more viewers a television show attracts, the greater the advertising revenue the show generates. The television network Vidnet's most popular show, *Starlight*, currently earns the network's highest profit, but next year, because of unavoidable increases in production costs, its profits are projected to fall to below the average for Vidnet shows. Therefore, Vidnet would earn greater profits overall if replaced *Starlight* with a show of average popularity and production costs.

Which of the following, if true, most seriously weakens the argument?

（A）The average profit of Vidnet shows has increased in each of the last three years.

（B）Shows that occupy time slots immediately before and after very popular shows tend to have far more viewers than they otherwise would.

（C）*Starlight* currently has the highest production costs of all Vidnet shows.

（D）Last year Vidnet lost money on a weekly show that was substantially similar to *Starlight* and was broadcast on a different day from *Starlight*.

（E）Even if, as a result of increased production costs, *Starlight* becomes less profitable than the average for Vidnet shows, it will still be more profitable than the average for television shows of all networks combined.

类别：普通预测推理

前提：用一个受欢迎程度和制作成本都一般的节目取代《星光大道》。

结论：Vidnet 将获得更大的整体利润。

推理：结论“Vidnet 将获得更大的整体利润”的充分条件为“新节目在总收入和总成本的差值上要高于《星光大道》”。答案选项需指出新节目在收入方面不如《星光大道》。

选项分析：

(A) 在过去三年中，Vidnet 节目的平均利润每年都在增加。过去的情况与讨论无关。

(B) 正确。占据非常受欢迎的节目前后的时间段的节目，其观众人数往往比其他节目要多得多。本选项指出了《星光大道》除了节目本身的利润之外的另一个利润点，即它可以带动其他节目。所以，如果取消《星光大道》，Vidnet 的总利润有可能减少。

(C) 在 Vidnet 的所有节目中，《星光大道》目前的制作成本最高。与讨论无关。

(D) 去年，Vidnet 因一档与《星光大道》基本相似的周播节目而亏损，该节目在与《星光大道》不同的日子播出。与讨论无关。

(E) 即使由于制作成本的增加，《星光大道》的利润低于 Vidnet 节目的平均水平，它仍然会比所有网络电视节目加起来的平均利润高。与所有网络电视节目加起来的平均利润无关。

7. Mangroves are salt-tolerant trees that form dense masses along sheltered tropical shorelines, including those on the island of Kalami. On Kalami, traditional uses of mangrove wood are quite limited, and no significant fishing goes on in the mangrove swamps. Therefore, the islanders would reap substantial economic gain by accepting a recent offer made by a mainland woodchip company to pay the islanders a fair price for the right to harvest the wood in their mangrove swamps.

Which of the following, if true, most seriously weakens the argument?

(A) Most of the species of fish that are important to the Kalami islanders' fishing industry depend on mangrove swamps as spawning and nursery areas.

(B) A significant portion of the money that Kalami islanders might receive from the wood chip company would be devoted to modernizing the island's fishing fleet.

(C) There is not enough indigenous forest left in the interior of Kalami to be of interest to the timber industry or even the woodchip industry.

(D) Mangroves do not thrive on the portion of the coastline of Kalami that is oriented toward the open ocean.

(E) Wood whose primary use is for woodchips is considerably less valuable than wood that is used in construction or in furniture making.

类别：普通预测推理

前提：在卡拉米岛，红树林木材的传统用途相当有限，而且在红树林沼泽中没有大量的捕鱼活动。

结论：如果接受一家大陆木片公司最近提出的向岛民支付合理价格以获得其在红树林沼泽中采伐木材的权力，岛民将获得巨大的经济收益。

推理：结论“岛民将获得巨大的经济收益”的充分条件应是红树林沼泽在任何方面都无法为村民提供经济收益。

选项分析：

(A) 正确。对卡拉米岛居民的捕鱼业很重要的大多数鱼种都依赖红树林沼泽作为产卵和繁殖区。红树林中的木材可能确实无用，但其中的沼泽对岛民的收入很有用。因此，如果失去沼泽，那么岛民们很可能蒙受经济损失。

(B) 卡拉米岛居民可能从木片公司得到的资金中，很大一部分将用于该岛捕鱼船队的现代化建设。

(C) 卡拉米岛内陆没有足够的本土森林，木材行业甚至是木片行业都不关注。

(D) 在卡拉米岛的海岸线上，红树林面向大海的部分并不繁茂。

(E) 主要用于制作木片的木材比用于建筑或家具制造的木材价值低得多。

8. Online shopping provides consumers with many product alternatives and a vast amount of product information. But purchasing common, everyday items—convenience goods—requires little brand comparison, planning, or new information. Consumers know exactly what to buy, and where to buy it, based on past shopping experience. So, in relation to convenience goods, online shopping has little advantage over traditional shopping.

The argument above is most vulnerable to the criticism that it

(A) fails to take into account that the most significant segment of online shopping sales is not convenience goods

(B) presumes, without providing justification, that convenience goods are typically products of a kind that everybody purchases at some time

(C) presumes, without providing justification, that online shopping has no significant advantages that are relevant besides product choice and product information

(D) fails to consider the advantages of traditional shopping in relation to certain categories of goods other than convenience goods

(E) fails to take into account that even ordinary, everyday items are sometimes not conveniently available to purchase

类别：泛化推理

前提：便利品几乎不需要进行品牌比较、规划或获取新信息（这些是网上购物的优点）。

结论：就便利品而言，网上购物与传统购物相比没有什么优势。

推理，品牌比较、规划或获取新信息无法代表所有优势或没有考虑网上购物的其他优势。

选项分析：

(A) 没有考虑到网上购物销售中最重要的部分不是便利品。

(B) 假设便利品通常是每个人都会在某些时候购买的产品。

(C) 正确。假设网上购物除了产品选择和产品信息外，没有其他相关的明显优势。

(D) 没有考虑到传统购物在便利品以外的某些类别商品上的优势。

(E) 没有考虑到即使是普通的日常用品，有时也不方便购买。

9. Most proposed legislation is introduced by legislators who, seeking to serve their own personal interests, stand to benefit if the legislation is passed into law. However, since most proposed legislation is not passed into law, most attempts by legislators to serve their own personal interests are unsuccessful.

The reasoning in the argument above is vulnerable to criticism on which of the following grounds?

(A) It does not take into account the fact that legislation can be proposed by people other than legislators.

(B) It does not acknowledge that some proposed legislation is introduced by legislators who are not attempting to serve their own personal interests.

(C) It ignores the possibility that proposed legislation that serves a legislator's personal interests might also serve the interests of other people.

(D) It neglects to indicate what percentage of proposed legislation introduced by legislators is passed into law.

(E) It falls to rule out explicitly the possibility that legislators successfully serve their own personal interests by means other than proposed legislation.

类别：泛化推理

前提：大多数拟议的立法都没有通过成为法律。

结论：立法者着眼于自己的个人利益的大多数尝试都是不成功的。

推理：指出拟议的立法没有通过成为法律不能代表所有的尝试都不成功，或者给出其他的能给立法者带来收益的方式。

选项分析：

(A) 它没有考虑到立法者以外的人也可以提出立法的事实。就算其他人提出立法，但只要这些拟议的法案不能成为法律，立法者的企图就是不成功的。

(B) 它不承认有些立法提案是由立法者提出的，而这些立法者并不是试图为他们自己的个人利益服务。与不打算满足自我利益的立法者无关。

(C) 它忽视了这样一种可能性，即为立法者个人利益服务的立法提案也可能为其他人的利益服务。立法提案服务于谁和讨论无关。

(D) 它忽略了指出立法者提出的立法提案中，有多大比例的提案通过成为法律。原文已经提到“大部分立法提案没能成为法律”。

(E) 正确。它没有明确排除立法者通过立法提案以外的手段成功地为自己的个人利益服务的可能性。

10. In Valgin, gasoline is sold in different grades, the most expensive being premium gasoline. Only about 10 percent of gasoline-powered vehicles sold in Valgin require premium gasoline, and for other vehicles premium provides no advantage. Nevertheless, nearly 15% of gasoline sold there is premium. So presumably a number of Valgian drivers are wasting money when they buy gasoline.

Which of the following, if true, most strengthens the argument?

(A) Buyers of vehicles that require premium gasoline do not drive substantially more miles per year than other drivers.

(B) More of the intermediate grade of gasoline is sold than of premium gasoline.

(C) Most of the vehicles that require premium gasoline are high-performance vehicles and are relatively expensive.

(D) Premium gasoline costs almost 25 percent more than least expensive grade of gasoline.

(E) Some very old vehicles that did not require premium gasoline when they were new now do require it to run smoothly.

类别：归因推理

前提：只有约 10% 的车需要高级汽油，但出售的汽油中有近 15% 是高级汽油。

结论：瓦尔金的一些司机在购买汽油时浪费钱。

推理：答案选项可以给出“瓦尔金的一些司机在购买汽油时浪费钱”的其他证据或给出其他能解释前提的原因。

选项分析

(A) 正确。购买需要高级汽油的车辆的人每年行驶的里程数并不比其他司机多。推理文段的前提中讲过了耗油量有 5% 的差距。本选项就排除了一个可能的原因，即好车开得多会更耗油。

(B) 中级汽油的销售量比高级汽油多。中级汽油的销量和讨论无关。

(C) 大多数需要高级汽油的车辆都是高性能车辆，而且价格相对较高。车的价格与讨论无关。

(D) 高级汽油的价格比最便宜的汽油高出近 25%。汽油价格具体相差多少与讨论无关。

(E) 一些非常老的车辆新买时不需要高级汽油，现在需要高级汽油才能顺利运行。本选项给出了另一个可能会导致“好车少但耗油量大”的原因，但问题问的是加强，本选项是削弱。

第三组

1. Environmentalist: Several thousand chemicals are found in our nation's drinking water, and the health effects of most of them are not completely known. Yet, in the past fifteen years, the government has added only one new chemical to the two dozen or so that it regulates in drinking water. Thus, the government's regulation of chemicals in drinking water is not sufficient to protect public health.

Finding the answer to which of the following questions is most likely to be the most helpful in evaluating the environmentalist's argument?

(A) How many of the chemicals the government currently regulates in drinking water have ever posed a risk to public health when found at high concentrations?

(B) By what means other than regulating chemicals in drinking water does the government work to protect public health?

(C) How many chemicals have been present in significant concentrations in the nation's drinking water for more than fifteen years while not being subject to government regulation?

(D) Why are health effects of most chemicals found in the nation's drinking water not completely known?

(E) Are there two or more chemicals that, at the concentrations present significantly often in the nation's drinking water over the last 15 years, can damage human health?

类别：普通预测推理

前提：在过去的 15 年里，政府在监管的饮用水中 20 多种化学品中只增加了一种新的化学品。

结论：政府对饮用水中的化学品的监管并不足以保护公众健康。

推理：结论“政府对饮用水中的化学品的监管并不足以保护公众健康”的充分条件应是“能威胁公众健康的饮用水不止新增一种化学品”。

选项分析：

(A) 政府目前监管的饮用水中的化学品，有多少在高浓度时曾对公众健康造成过风险？已经在监管范围内的化学品并不在我们讨论的范围内。

(B) 除了对饮用水中的化学品进行监管，政府还通过什么方式来保护公众健康？与其他方式无关。

(C) 有多少种化学品在全国的饮用水中以相当高的浓度存在了 15 年以上，却没有受到政府的监管？有多少种化学品没受到监管不重要，重要的是看其中“有害的”那些化学品是否被监管。

(D) 为什么在全国的饮用水中发现的大多数化学品对健康产生的影响并不完全清楚？与对健康产生的影响无关。

(E) 正确。是否有两种或更多的化学品，在过去 15 年里经常出现在全国的饮用水中，其浓度会损害人类健康？此选项讨论的是，新增的那一种是不是就是危害健康的全部化学品。如果是，那只增加一种足以保护公众健康；如果还有其他的有害化学品，那么只新增一种确实不足以保护公众健康。

2. For five years, every auto owner who has switched to Taske Auto Insurance has gotten a rate from 2 to 20 percent less than what they had been paying someone else for the same level of coverage. So clearly, if you are insured by someone else and you call Taske, you can be confident of this: The rate Taske offers you will save you at least 2 percent of what you currently pay!

Which of the following, if true, points to a flaw in the argument in the advertisement?

(A) Current Taske customers with the same level of automobile insurance coverage as each other sometimes pay different rates for that coverage.

(B) Of the people who switched to Taske, the vast majority achieved savings that were closer to 2 percent than to 20 percent.

(C) A large proportion of the auto owners who were quoted auto insurance rates by Taske chose not to switch their insurance to Taske.

(D) Some of Taske's competitors do not offer all of the different levels of coverage offered by Taske.

(E) Taske has been offering automobile insurance for only five years.

类别：泛化推理

前提：每一位转到 Taske 汽车保险的车主所获得的保险费率都比他们之前为同等水平的保险向别人支付的保险费低 2% ~20%。

结论：Taske 为您提供的费率将比您目前支付的费率节省至少 2%。

推理：转到 Taske 汽车保险的车主无法代表所有的人，或者其他人无法节省超过 2% 的费用。

选项分析：

(A) 目前拥有相同水平汽车保险的 Taske 公司的客户，有时会为该保险支付不同的费用。

(B) 在转到 Taske 的人中，绝大多数人节省了接近2%，而不是 20%。

(C) 正确。在 Taske 报出汽车保险费率的车主中，有很大一部分人选择不将他们的保险转到 Taske。如果 Taske 报出了 100 份保险，只有 5 人最终投保，那么，即便是这 5 人确实保费比较低，也不代表 Taske 给剩下的人报的价格也低。

(D) 一些 Taske 的竞争对手并没有提供 Taske 所提供的所有不同级别的保险。

(E) Taske 提供汽车保险只有 5 年的时间。

3. In Brindon County, virtually all of the fasteners—such as nuts, bolts, and screws—used by workshops and manufacturing firms have for several years been supplied by the Brindon Bolt Barn, a specialist wholesaler. In recent months many of Brindon County's workshops and manufacturing firms have

closed down, and no new ones have opened. Therefore, the Brindon Bolt Barn will undoubtedly show a sharp decline in sales volume and revenue for this year as compared to last year.

The argument depends on assuming which of the following?

(A) Last year the Brindon Bolt Barn's sales volume and revenue were significantly higher than they had been the previous year.

(B) The workshops and manufacturing firms that have remained open have a smaller volume of work to do this year than they did last year.

(C) Soon the Brindon Bolt Barn will no longer be the only significant supplier of fasteners to Brindon County's workshops.

(D) The Brindon Bolt Barn's operating expenses have not increased this year.

(E) The Brindon Bolt Barn is not a company that gets the great majority of its business from customers outside Brindon County.

类别：普通预测推理

前提：布林顿县的许多车间和制造公司都关闭了，且没有新的公司开业。

结论：与去年相比，布林顿螺栓库今年的销售量和收入无疑会出现大幅下降。

推理：结论“销售量和收入会出现大幅下降”的充分条件是布林顿螺栓库的客户绝大部分都是布林顿县的车间和制造公司，且这些车间和制造公司减少了。

选项分析：

(A) 去年布林顿螺栓库的销售量和收入明显高于前一年的水平。将去年和前年做比较没有意义。

(B) 继续营业的车间和制造公司今年的工作量比去年要小。由于大部分的车间和制造公司关闭，所以无论继续营业的公司的工作量增加还是减少，对布林顿螺栓库的影响都不大。

(C) 很快，布林顿螺栓库将不再是布林顿县车间唯一重要的紧固件供应商。无论有没有别的供应商，与结论无关。

(D) 布林顿螺栓库今年的运营费用没有增加。讨论的是收入，与运营费用（成本）没有关系。

(E) 正确。布林顿螺栓库不是一个从布林顿县以外的客户那里获得绝大部分业务的公司。排除了“布林顿螺栓库大部分的收入来自布林顿县以外的客户”的可能性，满足了结论的充分条件。

4. In the United States, many companies provide health care insurance for their employees. More and more of those companies are implementing programs promoting healthy lifestyles, programs that focus on factors associated with rising medical costs, such as high blood pressure. When successful, such programs help reduce the costs of health care, and thus of health insurance. In Canada, however, the government pays for health care. Clearly, therefore, Canadians companies have no financial interest to implement similar programs.

Which of the following, if true, most seriously weakens the argument?

(A) When employees miss work because of illness, their employer typically incurs expenses related to finding temporary substitutes for those employees.

(B) Many United States companies have found that certain programs promoting healthy lifestyles improve employees' health far more than other such programs do.

(C) Some companies introduce programs promoting healthy lifestyles without first analyzing whether, in this specific case, there will be a net cost savings.

(D) For people who seek employment with a company, the presence or absence of a program promoting healthy lifestyles is not generally significant consideration.

(E) The cost of specific medical treatments is generally lower in Canada than it is in the United States.

类别：普通预测推理

前提：在加拿大，政府为医疗服务买单。

结论：加拿大的公司对实施促进健康生活方式的项目不考虑经济因素。

推理：结论“不考虑经济因素”的充分条件应是实施促进健康生活方式的项目没有任何经济收益。

选项分析：

(A) 正确。员工因病缺勤时，他们的雇主通常会承担寻找这些员工的临时替代者产生的相关费用。本选项直接给出了公司的一项经济激励，即这些临时工的费用是公司需要承担的。所以，实施促进健康生活方式的项目可以帮公司省下雇临时工的钱。

(B) 许多美国公司发现，某些促进健康生活方式的项目对员工健康的改善远远超过其他此类项目。项目的比较与讨论无关。

(C) 有些公司在引入促进健康生活方式的项目时，没有首先分析在这种特定情况下是否会节省净成本。有无分析特定情况与讨论无关。

(D) 对于在公司求职的人来说，是否有促进健康生活方式的项目，一般来说不是重要的考虑因素。员工求职与讨论无关。

(E) 在加拿大，具体的医疗费用一般比美国低。国家间医疗费用的比较与讨论无关。

5. Solar radiation is a cause of skin cancer. In the United States, skin cancer is more common on the left side of the face than on the right. Some dermatologists hypothesize that this difference is due to drivers in the United States being directly exposed to solar radiation on the left side of the face more often than on the right, since cars in the United States have the driver's side window to the driver's left.

Which of the following, if true, most strongly supports the dermatologists' hypothesis?

(A) Many people in the United States who develop skin cancer do not regularly drive a car.

(B) When the windows of a car are closed, they absorb much of the harmful solar radiation to which drivers would otherwise be exposed.

(C) In countries where cars have the driver's side window to the driver's right, skin cancer is more common on the right side of the face.

(D) In the United States, most drivers who are diagnosed with skin cancer continue to drive after being diagnosed.

(E) Many people who spend more time driving than the average also spend more time out of doors than the average.

类别：因果推理

前提：美国汽车的驾驶员侧窗户在司机的左边；左脸患皮肤癌比右脸更常见。

结论：左脸患皮肤癌比右脸更常见是因为美国司机的左脸更经常地直接暴露在太阳辐射下。

推理：

纯粹巧合：指出“受太阳辐射”和“患皮肤癌”具备能产生因果关系的原理或“受太阳辐射”和“患皮肤癌”具备正相关的关系。

因果倒置：排除患皮肤癌的人更可能受太阳辐射的可能性。

他因导致结果：排除其他导致皮肤癌的因素。

选项分析：

(A) 在美国，许多患皮肤癌的人并不经常开车。

(B) 当汽车的窗户关闭时，它们吸收了许多有害的太阳辐射，否则司机就会暴露其中。

(C) 正确。在汽车的驾驶员侧窗户在司机的右边的国家，皮肤癌出现在右脸更常见。本选项显然指出了结论中两个事件并非“纯粹巧合”，即在其他情况下两者依然同时出现。

(D) 在美国，大多数被诊断患有皮肤癌的司机在被诊断后仍继续开车。

(E) 许多比一般人开车时间多的人也比一般人在户外的时间多。

6. Science reporter: In an experiment, scientists fed one group of mice an ordinary diet and fed another group a diet fortified with the microbes found in yogurt. The mice then explored a maze made up of tunnels and uncovered paths. The mice on the fortified diet ventured out of the tunnels twice as often as the other mice did, suggesting that the mice on the fortified diet felt less anxiety about being out in the open. This shows that the differences in the mice's diets affected their anxiety levels.

Which of the following, if true, would most strongly support the science reporter's conclusion?

(A) In addition to having the added microbes, the fortified diet differed from the ordinary diet by having significantly fewer calories.

(B) The microbes found in yogurt produce a chemical known to lower anxiety levels.

(C) In a second experiment conducted to check the results of the first, the diets of the two groups of mice were switched, and then the mice ran the maze again.

(D) Dietary changes can affect mice's sensitivity to light and thereby affect the strength of their preference for dark, enclosed spaces.

(E) The anxiety levels of mice affect their behavior in many ways that the scientists could have easily observed during the experiment.

类别：因果推理

前提：使用强化饮食的老鼠冒险走出隧道的次数是其他老鼠的两倍，这表明使用强化饮食的老鼠对户外活动的焦虑感较低。

结论：老鼠饮食的不同影响了它们的焦虑水平。

推理：

纯粹巧合：指出“不同饮食”和“焦虑水平”具备能产生因果关系的原理，或说明在其他研究中，“不同饮食”和“焦虑水平”依旧具备相关性。
因果倒置：排除焦虑水平不同的老鼠饮食不同的可能性。
他因导致结果：排除其他导致焦虑水平不同的因素。

选项分析：

(A) 除了添加微生物外，强化饮食与普通饮食的不同之处在于其热量明显减少。此选项涉及他因，无法加强结论。
(B) 正确。在酸奶中发现的微生物会产生一种降低焦虑水平的化学物质。本选项给出了饮食和焦虑之间具有因果关系的原理。
(C) 在为检验第一次实验结果而进行的第二次实验中，两组老鼠的饮食被调换，然后老鼠再次跑迷宫。此选项只讲了第二次实验的内容，没有讲最后的结果。所以，无法帮助我们评估结论。
(D) 饮食变化可以影响老鼠对光的敏感性，从而影响它们对黑暗、封闭空间的偏好程度。与对光的敏感性无关。
(E) 老鼠的焦虑程度在很多方面影响它们的行为，科学家在实验中可以很容易地观察到。与科学家是否能观察到无关。

7. Athletic coach: A recent survey of secondary school athletes revealed that those who use vitamins, protein powder, and other similar nutritional supplements were much more likely to be successful in athletics than those who do not. This is clear evidence that such supplements are indeed effective.
The athletic coach's argument is most vulnerable to criticism on which of the following grounds?

(A) It overlooks the possibility that athletes might not reveal anything that might be perceived as giving them an unfair advantage.

(B) It overlooks the possibility that athletes who are more likely to achieve athletic success even without nutritional supplements are more likely to use such supplements.

(C) It takes for granted that athletes engaged in team sports would be affected by supplements in the same way as competitors in nonteam sports.

(D) It fails to adequately address the possibility that among those who achieve athletic success, some might be more fit than others regardless of whether they take supplements or not.

(E) It fails to adequately address the possibility that what is true of athletes as a group might not be true of certain individual athletes.

类别：因果推理

前提：那些使用维生素、蛋白粉和其他类似营养补充剂的运动员比不使用的运动员更容易在田径运动中获得成功。

结论：营养补充剂确实有效（服用营养补充剂导致成功）。

推理：

纯粹巧合：指出“服用营养补充剂”和“成功”不具备能产生因果关系的原理，或说明在其他研究中，“服用营养补充剂”和“成功”不具备相关性。

因果倒置：本身更容易成功的人才喝营养补充剂。

他因导致结果：给出其他导致更容易成功的因素。

选项分析：

(A) 它忽略了运动员可能不会透露任何可能被认为是给他们带来不公平优势的事情。

(B) 正确。它忽略了这样一种可能性，即使没有营养补充剂，也更有可能取得运动成功的运动员更有可能使用这种补充剂。

(C) 它想当然地认为参加团队运动的运动员会像参加非团队运动的竞争者那样受到补充剂的影响。

(D) 它没有充分考虑到这样一种可能性，即在那些取得运动成功的人中，有些人可能比其他人更健康，不管他们是否服用营养补充剂。

(E) 它没有充分解决这样一种可能性，即适合运动员作为一个团队成员的情况可能并不适合某些运动员的个人情况。

8. Vitamin D is obtained directly from food and synthesized by means of exposure to sunlight. Among people who eat similar diets and live in areas with equal numbers of sunny days, those who are in poor health generally have lower levels of vitamin D in their blood than do those in good health.

Hypothesis: Poor health can affect the body's ability to utilize available vitamin D.

Which of the following, if true, most seriously undermines the hypothesis?

(A) People who are in poor health tend to have their blood tested more often than do people in good health.

(B) Some people in good health also have low levels of vitamin D in their blood.

(C) People who are in poor health tend to stay indoors more frequently.

(D) Experts disagree on the minimum level of vitamin D required to maintain a healthy body.

(E) People in poor health who eat a diet that is relatively low in vitamin D also generally have relatively low levels of vitamin D in their blood.

类别：因果推理

前提：健康状况不佳的人血液中的维生素 D 水平一般低于健康状况良好的人。

结论：健康状况不佳会影响身体利用现有维生素 D 的能力。

推理：

纯粹巧合：指出“健康状况不佳”和“维生素 D 水平”不具备能产生因果关系

的原理，或说明在其他研究中，"健康状况不佳" 和 "维生素 D 水平" 不再具备相关性。

因果倒置：本身维生素 D 水平低的人才更容易不健康。

他因导致结果：给出其他导致维生素 D 水平低的因素。

选项分析：

(A) 健康状况不佳的人往往比健康状况良好的人进行血液检测更频繁。与血液检测的频率无关。

(B) 一些健康状况良好的人血液中的维生素 D 水平也很低。即使健康的人维生素 D 水平低，也不影响不健康和维生素 D 水平低的因果关系。如果改成 "存在另一群不健康的人，但他们的维生素 D 水平高"，那更能质疑结论。

(C) 正确。健康状况不佳的人往往待在室内更多。本选项给出了另一个维生素 D 缺乏的原因（他因导致结果），即不是因为身体健康不好，而是因为不出去晒太阳。

(D) 专家们对维持健康身体所需的维生素 D 的最低水平意见不一。与专家意见无关。

(E) 健康状况不佳的人，如果饮食中的维生素 D 含量相对较低，他们血液中的维生素 D 含量一般也相对较低。原文已经明确表示饮食相似，所以再讨论饮食没有意义。

9. Fearing competition from stores that rent video games for home use, owners of stores that sell video games lobbied for protective legislation. Citing as a precedent legislation that postpones home film rentals until one year after general release to theaters, the video sellers proposed as an equitable solution a plan that would postpone rental of any video game until it had been available for sale for one year.

Which of the following, if true, would support an objection by owners of video rental stores that the fairness of the proposed legislation is not supported by the precedent cited?

(A) Although the daily rental fee for home use of films is generally the same as that for video games, the average rental period for a video game is longer than that for a film.

(B) Film rentals for home use tend to be particularly strong during the first year after a film has been made available for both sale and rental at video rental stores.

(C) Most people are interested in playing only the latest video games and tend to find video games that have been available for over a year unappealing whether they have played them before or not, whereas films maintain their appeal far longer after their release.

(D) People who rent video games tend to play them by themselves, whereas people who rent films invite friends and neighbors to watch.

(E) A slight decline in revenues from films that have been recently released in theaters has been attributed to the growing market for rentals of films for home use.

类别：类比推理

前提：立法规定：家庭电影租赁推迟到影院全面上映后一年。

结论：将任何视频游戏的租赁推迟到它可以销售后一年。

推理：答案选项需要指出“家庭电影租赁”和“视频游戏租赁”的其他相关相似点缺失。

选项分析：

(A) 尽管家庭电影的日租费一般与视频游戏的日租费相同，但视频游戏的平均租期比电影长。此选项虽然指出了两者的不同，但租期长短并不是“相关”相似点。我们应该找“推迟一年”这件事上两者的不同。

(B) 在一部影片在音像出租店同时出售和出租后的第一年，家庭电影的租赁往往特别强劲。此选项没有提到两者的不同。

(C) 正确。大多数人只对玩最新的视频游戏感兴趣，都会觉得已经超过一年的视频游戏不吸引人，不管他们以前是否玩过，而电影在上映后保持其吸引力的时间要长得多。此选项指出了两者相关相似点的缺失。即只有最新的游戏才吸引力，所以不能推迟一年才允许租赁，而电影无论上映后多久都有吸引力，所以可以推迟租赁。

(D) 租视频游戏的人往往自己玩，而租电影的人则邀请朋友和邻居观看。此选项同 A，虽然提到了两者的不同点，但使用者不同不是相关的相似点。

(E) 最近在影院上映的电影收入略有下降，这是因为家庭电影的租赁市场在增长。此选项没有提到两者的不同。

10. Newspaper article：Currently there are several prominent candidates for the position of governor of Province X. The prime minister has endorsed one of these candidates，Ann Shuh. The last candidate the prime minister endorsed for a provincial governorship won the position over strong rival candidates. Therefore，Ann Shuh will probably win the governorship of Province X.

The newspaper article's argument is most vulnerable to criticism on which of the following grounds?

(A) It asserts a sweeping generalization about a group on the basis of evidence regarding only a single，possibly unrepresentative，member of that group.

(B) It takes for granted that if Ann Shuh wins the governorship，the prime minister's endorsement will have been the main cause of her victory.

(C) It relies on a single point of similarity between two situations，without adequately addressing the possibility of relevant differences between those situations.

(D) It confuses a claim about a past event with a prediction about a similar，future event.

(E) It overlooks the possibility that the last candidate the prime minister endorsed for a provincial governorship could not have won the position without that endorsement.

类别：类比推理

前提：总理支持的上一位省长候选人在强大的竞争对手面前赢得了这个职位。

结论：（被总理支持的）安·舒很可能会赢得X省省长的职位。

推理：答案选项需要指出“上一位省长候选人”和“安·舒”的其他相关相似点缺失。

选项分析：

（A）它在只涉及一个群体中可能不具代表性的单一成员的证据的基础上，对该群体进行了全面的概括。此选项是针对“样本偏差”的描述。

（B）它想当然地认为，如果安·舒赢得省长职位，总理的支持将是她获胜的主要原因。文段主要进行了类比推理，而非猜测安·舒赢得职位的主要原因是什么。

（C）正确。它依赖于两种情况之间的单一相似点，而没有充分解决这些情况之间可能存在的相关差异。这是类比模式最大的漏洞。

（D）它混淆了对过去事件的主张和对未来类似事件的预测。这里没有“混淆”两个事件，而是认为两个事件可以类比。

（E）它忽略了这样一种可能性，即如果没有总理的支持，上一位省长候选人不可能赢得这个职位。与上一位候选人赢得职位的原因无关。

2.4 方案（Plan）

方案类考题要求我们去构造或者评估一个方案。方案类考题在问题的用词上和“构建论证”以及“评估论证”大体相同，只不过都会与方案、策略和政策等词相关，例如：

(1) Which of the following, if true, casts the most serious doubt on the argument made for the proposal?

(2) Which of the following, if true, most seriously calls into question the advisability of implementing the proposal?

(3) Which of the following, if true, most strongly indicates that the legislators' proposal will fail to achieve its goal?

(4) Which of the following, if true, most helps to explain the apparent inconsistency in the results of the library's policy?

在解题策略上，这类考题也与“构建论证”或“评估论证”大体一致。我们需要时刻关注方案是否达到预设目的，例如：

前提：私家汽车是主要的空气污染源之一。我们应该减少空气污染。

结论：城市应该禁止所有的私家汽车，只允许公共交通。

一个完整的方案应包括方案本身和方案目标。我们可以将其重构为：

目标：减少空气污染。

方案：禁止所有的私家汽车，只允许公共交通。

请注意，在方案尚未开始的时候，结论是方案的目标；在方案开始生效后，结论大概率将是方案的结果（效果）。

方案类推理和假说论证具有类似的评估维度：

（1）方案本身能否落地执行：可能出于某种原因，无法禁止私家汽车而仅允许公共交通的使用。

（2）目标是否能被方案实现：这个维度主要是评估方案是否能达成目标。

此外，因为在方案成功实施后，方案与目标经常可以转化为因和果的关系（方案是因，目标是结果），所以有时候方案也可以和因果推理结合，例如：

前提：凡是努力学习的人一般成绩都比较好。

结论：为了提高成绩，你应该努力学习。

上述例题中方案的制订基于“努力学习导致成绩提高”这一基本的因果关系，因此，这道题应该用因果推理的评估维度进行解题。

例题 1

Trancorp currently transports all its goods to Burland Island by truck. The only bridge over the channel separating Burland from the mainland is congested, and trucks typically spend hours in traffic. Trains can reach the channel more quickly than trucks, and freight cars can be transported to Burland by barges that typically cross the channel in an hour. Therefore, to reduce shipping time, Trancorp plans to switch to trains and barges to transport goods to Burland.

Which of the following, if true, casts most serious doubt on whether Trancorp's plan will succeed?

(A) It does not cost significantly more to transport goods to Burland by truck than it does to transport goods by train and barge.

(B) The number of cars traveling over the bridge into Burland is likely to increase slightly over the next two years.

(C) Because there has been so much traffic on the roads leading to the bridge between Burland and the mainland, these roads are in extremely poor condition.

(D) Barges that arrive at Burland typically wait several hours for their turn to be unloaded.

(E) Most trucks transporting goods into Burland return to the mainland empty.

类别：方案

目标：减少运输时间。

方案：Trancorp 打算用火车和驳船来向 Burland 运输物资。

推理：答案选项需指出“Trancorp 无法使用火车和驳船来向 Burland 运输物资”或“火车和驳船不能减少运输时间”。

选项分析

(A) 用卡车送货到 Burland 不会比用火车或者驳船送货到 Burland 贵很多。如果火车和驳船比卡车贵很多，这是在一定程度上刻意给出火车和驳船的“问题”。但本选项讨论的是卡车是否更糟糕，和方案没关系。

(B) 在未来两年内，通过大桥到达 Burland 的汽车数量将会上升。汽车数量上升和方案无关。

(C) 因为通往 Burland 和大陆之间大桥的道路交通繁忙，所以这些道路的状况都很差。本选项解释了汽车为什么会速度慢，但是和方案无关。

(D) 正确。到达 Burland 的驳船需要等很长一段时间才能卸货。如果本选项成立，则方案将无法达成“减少运输时间”的目标。

(E) 大部分送货到 Burland 的卡车都会空车返回。卡车是否会空车返回和方案无关，可以排除。

例题2

The economy around Lake Paqua depends on fishing of the lake's landlocked salmon population. In recent years, scarcity of food for salmon there has caused a decline in both the number and the size of the adult salmon in the lake. As a result, the region's revenues from salmon fishing have declined significantly. To remedy this situation, officials plan to introduce shrimp, which can serve as a food source for adult salmon, into Lake Paqua.

Which of the following, if true, most seriously calls into question the plan's chances for success?

(A) Salmon is not a popular food among residents of the Lake Paqua region.

(B) Tourists coming to fish for sport generate more income for residents of the Lake Paqua region than does commercial fishing.

(C) The shrimp to be introduced into Lake Paqua are of a variety that is too small to be harvested for human consumption.

(D) The primary food for both shrimp and juvenile salmon is plankton, which is not abundant in Lake Paqua.

(E) Fishing regulations prohibit people from keeping any salmon they have caught in Lake Paqua that is smaller than a certain minimum size.

类别：方案

目标：解决这种情况（让成年三文鱼的重量和数量上升，从而增加收入）。

方案：引入虾。

推理：答案选项需指出“无法引入虾”或“虾不能使成年三文鱼的重量和数量上升”。

选项分析

(A) 在 Lake Paqua 地区的居民看来，三文鱼并不是一种受欢迎的食物。本选项没有讨论方案的问题，可以排除。

(B) 对于 Lake Paqua 地区的居民来说，游客的钓鱼活动带来的收入比商业捕鱼带来的收入多。本选项没有讨论方案的问题，可以排除。

(C) 被引入的虾太小了，以至于不会被人类所捕食。引入虾是为了给鱼吃的，这和人是否会吃无关。

(D) 正确。虾和幼年三文鱼的主要食物都是湖里的蜉蝣，而蜉蝣并不是很多。如果虾会和幼年三文鱼抢食物，那么，从长期来看，三文鱼的数量肯定会下降（幼年三文鱼减少）。本选项指出了方案无法达成目标。

(E) 捕鱼规定要求人们不能带走比一个特定最小值更小的三文鱼。本选项没有讨论方案的问题，可以排除。

例题 3

In parts of South America, vitamin A deficiency is a serious health problem, especially among children. In one region, agriculturists hope to improve nutrition by encouraging farmers to plant a new variety of sweet potato called SPK004 that is rich in beta-carotene, which the body converts into vitamin A. The plan has good chances of success, since sweet potato is a staple of the region's diet and agriculture, and the varieties currently grown contain little beta-carotene.

Which of the following, if true, most strongly supports the prediction that the plan will succeed?

(A) There are other vegetables currently grown in the region that contain more beta-carotene than the currently cultivated varieties of sweet potato do.

(B) The flesh of SPK004 differs from that of the currently cultivated sweet potatoes in color and texture, so traditional foods would look somewhat different when prepared from SPK004.

(C) For successful cultivation of SPK004, a soil significantly richer in nitrogen is needed than is needed for the varieties of sweet potato currently cultivated in the region.

(D) There are no other varieties of sweet potato that are significantly richer in beta-carotene than SPK004 is.

(E) The currently cultivated varieties of sweet potato contain no important nutrients that SPK004 lacks.

类别：方案

目标：增加营养物质的摄入。

方案：引入 SKP004。

推理：答案选项需指出“无法引入 SKP004”或“引入 SKP004 无法增加营养物质的摄入”。

选项分析

(A) 在该地区，没有其他蔬菜比现在种的红薯含有更多的 β-胡萝卜素。无论其他蔬菜怎么样，只要 SPK004 这种红薯有好处，就可以得到现有的结论。所以本选项不能评估方案推理。

(B) SPK004 果肉的颜色和质地与现在种植的红薯不同，所以用 SPK004 做出来的传统食物会和现在看起来不一样。无论做出的东西和现在是不是一样的，只要大家依然会吃这种食物，那么就不会给方案带来什么问题。

(C) 为了成功种植 SPK004，其土壤的含氮量需要比现在种植红薯的土壤的含氮量高。本选项给方案增加了额外的困难，但是其不能证明这个内容可以使方案不可行，可以排除。

(D) 没有其他种类的红薯比 SPK004 更富含 β-胡萝卜素。此选项只讲了含有 β-胡萝卜素的量，但其含量不是评价经济效益的标准（我们不知道其他种类的红薯需要的人力和物力怎么样），所以就算有其他的红薯含有更多有效成分，最多可以质疑我们是不是要选取种植 SPK004 这个东西，但是并不能反驳种植 SPK004 这个方案是否可以让人们增加营养的摄入。

(E) 正确。现在种植的红薯不含 SPK004 缺少的重要营养物质。如果现在的红薯含有 SPK004 缺少的东西，那么将现在的红薯全部换成 SPK004 可能就会无法达成“增加营养”这个目标。

例题 4

In the nation of Partoria, large trucks currently account for 6 percent of miles driven on Partoria's roads but are involved in 12 percent of all highway fatalities. The very largest trucks—those with three trailers—had less than a third of the accident rate of single-and double-trailer trucks. Clearly, therefore, one way for Partoria to reduce highway deaths would be to require shippers to increase their use of triple-trailer trucks.

Which of the following, if true, most seriously weakens the argument?

(A) Partorian trucking companies have so far used triple-trailer trucks on lightly traveled sections of major highways only.

(B) No matter what changes Partoria makes in the regulation of trucking, it will have to keep some smaller roads off-limits to all large trucks.

(C) Very few fatal collisions involving trucks in Partoria are collisions between two trucks.

(D) In Partoria, the safety record of the trucking industry as a whole has improved slightly over the past ten years.

(E) In Partoria, the maximum legal payload of a triple-trailer truck is less than three times the maximum legal payload of the largest of the single-trailer trucks.

类别：因果推理

前提："三个车厢的卡车"和"事故率更低"之间存在正相关关系。

结论：为了使事故率更低，用三个车厢的卡车吧。

推理：

纯粹巧合（削弱）：指出"三个车厢的卡车"和"事故率更低"之间不存在能产生因果关系的原理，或说明在其他场景中，"三个车厢的卡车"和"事故率更低"不再具备正相关的关系。

因果倒置（削弱）：指出那些本身就事故率更低的卡车才用三个车厢。

他因导致结果（削弱）：给出其他可能导致事故率更低的因素，比如，开三厢卡车的司机都是最谨慎的，是司机的原因才导致事故率低。

选项分析

(A) 正确。Partoria 的卡车公司目前都将三厢卡车派往车流量最少的公路上行驶。本选项提出了一个他因可以导致事故率更低。

(B) 无论 Partoria 对卡车的限制如何变化，Partoria 仍然会让一些比较小的路不能通过大型卡车。本选项与讨论无关。

(C) 很少有事故是在两辆卡车之间发生的。本选项讨论的是事故发生的原理。

(D) 在 Partoria，卡车工业的安全记录在过去十年里上升了。本选项讨论的是卡车工业整体的情况，和讨论无关。

(E) 在 Partoria，三厢卡车的载货量少于一厢卡车的三倍。卡车的载货量与讨论无关。

练 习

1. Much of the materials in landfills is paper and other organic materials that decompose gradually. Since these materials decompose much faster when wet, keeping landfills wet would free up space in them faster, thereby both reducing the need for new landfills and slowing expansion of existing landfills. Northwest Environmental Commission is considering a proposal to install watering systems in landfills with the aim of reducing the landfills' adverse environmental impact on the region.

Which of the following, if true, most strongly supports the prediction that the proposal, if adopted, will not have the intended effect?

(A) Organic material that is kept wet decomposes even faster, if it is also kept warm.

(B) Installing watering systems in landfills would add significantly to the costs of landfill management.

(C) Keeping landfills wet increases the risk that polluted water will leak out into streams, rivers, and lakes.

(D) Paper and cardboard that are wet are more easily compressed and therefore take up less space.

(E) Some of the materials in landfills do not decompose more rapidly when exposed to moisture.

2. Sales manager: My salespeople are required to attend classes to keep their sales skills current. We use a lot of company time organizing this training—renting meeting rooms, hiring trainers, and holding sessions during business hours. We can decrease the amount of company time devoted to salespeople's training by requiring the sales staff to use online tutorials and videos instead.

Which of the following, if true, most strongly supports the idea that the sales manager's proposal will have the predicted effect?

(A) Studies have shown that supplementing classroom activities with online materials can greatly increase peoples' ability to learn the subject matter.

(B) The online tutorials and videos can be tailored to address issues specific to the sales manager's industry.

(C) Employees will be more willing to take the sales classes if they can do so at their own pace.

(D) The sales staff are likely to access the online classes and videos during business hours.

(E) Online tutorials and videos once created require almost no work to maintain.

3. The Acme Corporation plans to improve productivity by giving staff more powerful computers with faster Internet connections. But some of the corporation's managers claim that many employees play games over the Internet at work, and conclude from this claim that the plan will be counterproductive.

Assuming that the managers' claim is correct, which of the following, if true, would most strongly support the prediction that the Acme Corporation's plan will succeed?

(A) The Acme Corporation has a written policy forbidding staff from playing computer games at work.

(B) None of the managers at Acme Corporation play games over the Internet at work.

(C) At most corporations, there is no correlation between the amount of time employees spend playing computer games and the power of their computers or speed of their Internet connections.

(D) Few if any employees at Acme Corporation have duties that could be performed more efficiently with more powerful computers or faster Internet connections.

(E) Managers at Acme Corporation have observed that those employees with the more powerful computers arid fast Internet connections spend a significant amount of time playing computer games.

4. Chickens raised on feed treated with compound Q gained an average of 5 percent more weight than chickens raised on untreated feed. Since chickens are sold by weight, farmers plan to treat all their feed with compound Q in an attempt to increase profits.

Which of the following provides the strongest evidence that the plan's objectives might not be reached?

(A) Chickens raised for egg production whose feed is treated with compound Q lay more eggs than do chickens whose feed is not so treated.

(B) Chicken feed must be treated with compound Q at least one week before the feed is given to the chickens.

(C) Compound Q has been known to help combat avian diseases that can destroy an entire flock of chickens.

(D) Treating chicken feed with compound Q raises the cost of chicken feed by 40 Percent.

(E) Compound Q turns feed blue, a color to which chickens are particularly attracted.

5. Medical research shows that recovering hospital patients heal better when they have more visitors, and when they can look out of their windows and see trees and parks rather than concrete and cars. In order to facilitate patient recovery,

the Smithville hospital is therefore planning a new wing with innovations such as single rooms, a landscaped park, reduced-rate underground parking, and improved dining facilities for visitors.

Which of the following, if true, would most increase the chances of the plan's helping patients recover?

(A) Hospital occupancy rates are somewhat higher for single rooms than for multiple-bed rooms.

(B) A number of agencies provide free art materials and lessons for hospital patients in recovery.

(C) Friends and relatives are more likely to visit patients in single rooms than in multiple-bed rooms.

(D) If patients have slightly longer hospital stays, the extra cost of creating a pleasing environment can be recovered in one year.

(E) The quality of hospital food has been shown to have an effect on healing rates in patients.

6. Spartina alterniflora, a grass native to salt marshes in eastern North America, was accidentally introduced into the continent's West Coast. Now S. alterniflora threatens to displace a closely related West Coast grass, Spartina foliosa. The invasion of an alien plant can sometimes be halted by introducing into the invaded area an insect that eats the invading species. Conservationists are considering this strategy as a way of combating further displacement of S. foliosa by S. alterniflora.

Determining which of the following would be most useful in assessing the strategy under consideration by the conservationists?

(A) Whether any insects were accidentally introduced into the West Coast when S. alterniflora was accidentally introduced there

(B) Whether S. alterniflora's displacement of S. foliosa is being hastened by insects that feed on S. foliosa but not on S. alterniflora

(C) Whether S. alterniflora has ever been saved from the invasion of an alien species by the introduction of an insect

(D) Whether there are insects that, if introduced into the invaded area, would eat only S. alterniflora

(E) Whether, if the strategy under consideration were to halt the displacement of S. foliosa by S. alterniflora, S. foliosa would regain the territory it has already lost

7. To combat forest fires in Country X, the forestry ministry has been clearing susceptible areas of small trees; but this is expensive, and the resulting small logs often go to waste because there is little demand for them. The ministry therefore plans to encourage local businesses to invest in processes to convert the logs into potentially valuable products, such as flooring or wood chips for heating, and to subsidize the businesses in the purchase of new equipment. The hope is that if these businesses become viable, sales of the small logs will consistently help offset the ministry's costs.

Which of the following, if true, would most increase the plan's chances of success?

(A) The forestry ministry increased its budget for forest thinning by 15 percent over the last six years.

(B) If a large multinational wood-processing corporation can be attracted to Country X, it would help to lower the rate of unemployment.

(C) The forestry ministry plans to define carefully the size specifications of the logs in question.

(D) The types of wood in the small trees taken from Country X forests are not all suitable as raw material for flooring or wood chips.

(E) A typical building in country X can be heated with reasonably priced wood chips for only about 20 percent of the cost of using conventional fuel.

8. The Internet service provider CableWeb is spending billions of dollars to expand its network to meet the growing demand for Internet service. CableWeb cannot raise subscription rates to pay for this expansion, or else many of its subscribers will switch to competing service providers. Thus, to generate funds for this expansion, CableWeb plans to require payments from owners of commercial Websites each time a CableWeb subscriber accesses those sites.

In order to evaluate how likely CableWeb's planned required payments will be to generate funds for expansion, it would be most helpful to know which of the following?

(A) To what extent owners of commercial Websites depend on revenues generated from consumers who access their sites via CableWeb's services
(B) How many Internet service providers competing with CableWeb are also expanding their networks to meet the growing demand for Internet service
(C) By what methods other than demanding payments from owners of commercial Websites CableWeb could feasibly pay to expand its network
(D) How CableWeb's subscription rates currently compare with those of its main competitors
(E) Whether any CableWeb subscribers would be willing to pay a fee each time they access a commercial Website

9. Career counselor: Research has found that people who hunt extensively for the highest-paying careers earn, on average, about 20 percent more than other people. But they also experience less job satisfaction and more anxiety at work.

This suggests that to find a fulfilling career, it is best not to focus primarily on maximizing your salary.

Which of the following, if true, would most weaken the career counselor's argument?

(A) People who find their careers fulfilling generally report higher job satisfaction and less anxiety at work than do people who find their careers unfulfilling.

(B) Most people who report that their careers are exceptionally fulfilling earn relatively modest salaries.

(C) On average, people who tend in general to feel less satisfaction and more anxiety are more likely to hunt extensively for the highest-paying careers.

(D) On average, people who hunt extensively for the most fulfilling careers, with little regard for salary, still earn more than other people.

(E) In some situations, it is much more important to find a career offering a higher salary than a career that is fulfilling.

10. Journalist: Researchers surveyed 1,000 people who had experienced personal catastrophes. They found that those who formulated new personal goals after such disasters achieved personal readjustment more effectively and more quickly than those who had not done so. Assuming that the study was done correctly, it follows that to bring about readjustment in such people most quickly and most effectively, counselors should mainly focus on encouraging their clients to formulate new personal goals.

Which of the following most accurately describes a flaw in the journalist's reasoning?

(A) The journalist relies on the use of emotively colored language, rather than on the presentation of data.

(B) The argument fails to consider that the readjustment even of those who have not experienced personal catastrophes might be expedited by the formulation of new personal goals.

(C) The argument fails to consider whether, for those who have experienced personal catastrophes in their lives, maintaining previous personal goals might be beneficial for readjustment.

(D) The journalist's generalization about what counselors should do relies on a study that, in its collection of data, did not include a sample of counselors.

(E) The argument fails to consider the possibility that formation of new personal goals is an early stage of readjustment rather than its cause.

答案及解析

1. Much of the materials in landfills is paper and other organic materials that decompose gradually. Since these materials decompose much faster when wet, keeping landfills wet would free up space in them faster, thereby both reducing the need for new landfills and slowing expansion of existing landfills. Northwest Environmental Commission is considering a proposal to install watering systems in landfills with the aim of reducing the landfills' adverse environmental impact on the region.

Which of the following, if true, most strongly supports the prediction that the proposal, if adopted, will not have the intended effect?

(A) Organic material that is kept wet decomposes even faster, if it is also kept warm.

(B) Installing watering systems in landfills would add significantly to the costs of landfill management.

(C) Keeping landfills wet increases the risk that polluted water will leak out into streams, rivers, and lakes.

(D) Paper and cardboard that are wet are more easily compressed and therefore take up less space.

(E) Some of the materials in landfills do not decompose more rapidly when exposed to moisture.

类别：方案

目标：减少垃圾填埋场对该地区环境的不利影响。

方案：在垃圾填埋场安装浇灌系统。

推理：答案选项需指出“无法在垃圾填埋场安装浇灌系统”或“在垃圾填埋场安装浇灌系统无法真的减少不利影响”。

选项分析：

(A) 保持湿润的有机物，如果同时保持温暖，则分解得更快。本选项可以加强推理。

(B) 在垃圾填埋场安装浇灌系统会大大增加垃圾填埋场的管理成本。本选项给出的是操作成本会增加，而不是“无法操作”，所以不能评估推理。

(C) 正确。保持垃圾填埋场湿润会增加被污染的水渗入溪流、河流和湖泊的风险。本选项显然指出了在垃圾填埋场安装浇灌系统最终无法减少不利影响。

(D) 湿的纸张和硬纸板更容易被压缩，因此占用的空间更小。与占用空间无关。

(E) 垃圾填埋场中的一些材料暴露于湿气中时不会分解得更快。只要能分解即可，分解得快慢与讨论无关。

2. Sales manager: My salespeople are required to attend classes to keep their sales skills current. We use a lot of company time organizing this training—renting meeting rooms, hiring trainers, and holding sessions during business hours. We can decrease the amount of company time devoted to salespeople's training by requiring the sales staff to use online tutorials and videos instead.

Which of the following, if true, most strongly supports the idea that the sales manager's proposal will have the predicted effect?

(A) Studies have shown that supplementing classroom activities with online materials can greatly increase peoples' ability to learn the subject matter.

(B) The online tutorials and videos can be tailored to address issues specific to the sales manager's industry.

(C) Employees will be more willing to take the sales classes if they can do so at their own pace.

(D) The sales staff are likely to access the online classes and videos during business hours.

(E) Online tutorials and videos once created require almost no work to maintain.

类别：方案

目标：减少公司用于培训销售人员的时间。

方案：销售人员使用在线教程和视频。

推理：答案选项需指出“销售人员可以使用在线教程和视频”或“使用在线教程和视频可以真的减少公司用于培训销售人员的时间”。

选项分析：

(A) 研究表明，用在线材料来补充课堂活动可以大大提高人们学习主题内容的能力。目标是减少时间，与能否提高学习能力无关。

(B) 在线教程和视频可以定制，以解决销售经理所在行业的具体问题。在线教程和视频的内容与讨论的方案和目标无关。

(C) 如果员工可以按照自己的节奏学习，他们会更愿意参加销售课程。与讨论的方案和目标无关。

(D) 销售人员有可能在工作时间访问在线课程和视频。销售人员在何时上课与讨论的方案和目标无关。

(E) 正确。在线教程和视频一旦创建，几乎不需要任何维护工作。如果可以通过要求销售人员使用在线教程和视频来减少公司用于培训销售人员的时间，那么我们对于在线教程和视频的维护时间一定不能超过其他形式的培训时间，否则方案将无法达成目标。

3. The Acme Corporation plans to improve productivity by giving staff more powerful computers with faster Internet connections. But some of the corporation's managers claim that many employees play games over the Internet at work, and conclude from this claim that the plan will be counterproductive.

Assuming that the managers' claim is correct, which of the following, if true, would most strongly support the prediction that the Acme Corporation's plan will succeed?

(A) The Acme Corporation has a written policy forbidding staff from playing computer games at work.

(B) None of the managers at Acme Corporation play games over the Internet at work.

(C) At most corporations, there is no correlation between the amount of time employees spend playing computer games and the power of their computers or speed of their Internet connections.

(D) Few if any employees at Acme Corporation have duties that could be performed more efficiently with more powerful computers or faster Internet connections.

(E) Managers at Acme Corporation have observed that those employees with the more powerful computers arid fast Internet connections spend a significant amount of time playing computer games.

类别：方案

目标：提高产量。

方案：给员工提供更强大的电脑和更快的互联网连接。

推理：答案选项需指出"可以给员工提供更强大的电脑和更快的互联网连接"或"给员工提供更强大的电脑和更快的互联网连接可以真的提高产量"。

选项分析：

(A) Acme 公司有一项书面政策，禁止员工在工作时玩电脑游戏。员工在工作时玩游戏是既定事实，无论有无书面政策，都与讨论无关。

(B) Acme 公司的经理们没有一个人在工作时间在网上玩游戏。经理是否以身作则与讨论无关。

（C）正确。在大多数公司，员工玩电脑游戏的时间与他们电脑的功率或互联网连接的速度没有关联。本选项显然指出了玩游戏和电脑的好坏没有关联，相当于支持了方案可以达成目标。

（D）在 Acme 公司，很少有员工可以通过更强大的电脑或更快的互联网连接来更有效地履行职责。此选项指出方案可能没用，削弱了方案达成目标的可能性。

（E）Acme 公司的经理观察到，那些拥有更强大的电脑和快速互联网连接的员工花了大量时间玩电脑游戏。本选项可以算作对方案的削弱。

4. Chickens raised on feed treated with compound Q gained an average of 5 percent more weight than chickens raised on untreated feed. Since chickens are sold by weight, farmers plan to treat all their feed with compound Q in an attempt to increase profits.

Which of the following provides the strongest evidence that the plan's objectives might not be reached?

(A) Chickens raised for egg production whose feed is treated with compound Q lay more eggs than do chickens whose feed is not so treated.

(B) Chicken feed must be treated with compound Q at least one week before the feed is given to the chickens.

(C) Compound Q has been known to help combat avian diseases that can destroy an entire flock of chickens.

(D) Treating chicken feed with compound Q raises the cost of chicken feed by 40 Percent.

(E) Compound Q turns feed blue, a color to which chickens are particularly attracted.

类别：方案

目标：增加利润。

方案：用化合物 Q 处理他们所有的饲料。

推理：答案选项需指出“无法用化合物 Q 处理他们所有的饲料”或“用化合物 Q 处理他们所有的饲料无法增加利润”。

选项分析：

(A) 为生产鸡蛋而饲养的鸡，其饲料经过化合物 Q 的处理，比饲料没有经过处理的鸡产下更多的蛋。

(B) 鸡饲料必须在给鸡喂食前至少一周用化合物 Q 处理。

(C) 已知化合物 Q 有助于防治可摧毁整个鸡群的禽类疾病。

(D) 正确。用化合物 Q 处理鸡饲料使鸡饲料的成本提高了 40%。成本增加，自然利润可能减少，也就是方案无法达成目标。

(E) 化合物 Q 使饲料变成蓝色，这是鸡特别喜欢的一种颜色。

5. Medical research shows that recovering hospital patients heal better when they have more visitors, and when they can look out of their windows and see trees and parks rather than concrete and cars. In order to facilitate patient recovery, the Smithville hospital is therefore planning a new wing with innovations such as single rooms, a landscaped park, reduced-rate underground parking, and improved dining facilities for visitors.

Which of the following, if true, would most increase the chances of the plan's helping patients recover?

(A) Hospital occupancy rates are somewhat higher for single rooms than for multiple-bed rooms.

(B) A number of agencies provide free art materials and lessons for hospital patients in recovery.

(C) Friends and relatives are more likely to visit patients in single rooms than in multiple-bed rooms.

(D) If patients have slightly longer hospital stays, the extra cost of creating a pleasing environment can be recovered in one year.

(E) The quality of hospital food has been shown to have an effect on healing rates in patients.

类别：方案

目标：促进病人的康复。

方案：史密斯维尔医院正在规划一个新的侧翼部分，其创新之处包括：单人房、景观公园、降低收费的地下停车场，以及为访客提供更好的餐饮设施。

推理：答案选项需指出“可以实现此创新”或“此创新可以真的促进病人的康复”。

选项分析：

(A) 医院单人病房的入住率比多人病房的入住率高一些。

(B) 许多机构为康复中的病人提供免费的艺术材料和课程。

(C) 正确。朋友和亲戚更可能探望单人病房的病人，而不是多人病房的病人。本选项给出了方案可以达成目标。因为题干中已经写明，更多的探望者会有助于病人的康复。因此，如果单人病房能带来更多的探望者，那么肯定可以促进病人的康复。

(D) 如果病人住院时间更长一点，创造愉悦环境的额外成本可以在一年内收回。

(E) 医院食物的质量已被证明对病人的痊愈率有影响。

6. Spartina alterniflora, a grass native to salt marshes in eastern North America, was accidentally introduced into the continent's West Coast. Now S. alterniflora threatens to displace a closely related West Coast grass, Spartina foliosa. The invasion of an alien plant can sometimes be halted by introducing into the invaded area an insect that eats the invading species. Conservationists are considering this strategy as a way of combating further displacement of S. foliosa by S. alterniflora.

Determining which of the following would be most useful in assessing the strategy under consideration by the conservationists?

(A) Whether any insects were accidentally introduced into the West Coast when S. alterniflora was accidentally introduced there

(B) Whether S. alterniflora's displacement of S. foliosa is being hastened by insects that feed on S. foliosa but not on S. alterniflora

(C) Whether S. alterniflora has ever been saved from the invasion of an alien species by the introduction of an insect

(D) Whether there are insects that, if introduced into the invaded area, would eat only S. alterniflora

(E) Whether, if the strategy under consideration were to halt the displacement of S. foliosa by S. alterniflora, S. foliosa would regain the territory it has already lost

类别：方案

目标：防止 S. foliosa 被 S. alterniflora 进一步取代。

方案：在被入侵地区引入一种吃入侵植物的昆虫。

推理：答案选项需指出“能否在被入侵地区引入一种吃入侵植物的昆虫”或“在被入侵地区引入一种吃入侵植物的昆虫能否真的防止 S. foliosa 被 S. alterniflora 进一步取代”。

选项分析：

(A) 当 S. alterniflora 被意外地引入西海岸时，是否有任何昆虫被意外地引入那里。与其他昆虫无关。

(B) S. alterniflora 是否被以 S. foliosa 为食但不以 S. alterniflora 为食的昆虫是否加速了 S. alterniflora 取代 S. foliosa。与别的昆虫无关。

(C) S. alterniflora 是否曾因引入昆虫而免遭外来物种的入侵。与 S. alterniflora 曾经是否被拯救过无关。

(D) 正确。是否有一些昆虫如果被引入被入侵地区，只吃 S. alterniflora。仅当昆虫只吃 S. alterniflora 而不吃 S. foliosa 时，方案才可以达成目标，且不会带来更多问题。

(F) 如果考虑的策略是阻止 S. foliosa 被 S. alterniflora 所取代，S. foliosa 是否会重新获得它已经失去的领土。与成功阻止之后，S. foliosa 能否恢复无关。

7. To combat forest fires in Country X, the forestry ministry has been clearing susceptible areas of small trees; but this is expensive, and the resulting small logs often go to waste because there is little demand for them. The ministry therefore plans to encourage local businesses to invest in processes to convert the logs into potentially valuable products, such as flooring or wood chips for heating, and to subsidize the businesses in the purchase of new equipment. The hope is that if these businesses become viable, sales of the small logs will consistently help offset the ministry's costs.

Which of the following, if true, would most increase the plan's chances of success?

(A) The forestry ministry increased its budget for forest thinning by 15 percent over the last six years.

(B) If a large multinational wood-processing corporation can be attracted to Country X, it would help to lower the rate of unemployment.

(C) The forestry ministry plans to define carefully the size specifications of the logs in question.

(D) The types of wood in the small trees taken from Country X forests are not all suitable as raw material for flooring or wood chips.

(E) A typical building in Country X can be heated with reasonably priced wood chips for only about 20 percent of the cost of using conventional fuel.

类别：方案

目标：小原木可以产生销售收入。

方案：鼓励当地企业进行投资，将原木转化为有潜在价值的产品，如地板或供暖用的木屑，并为企业购买新设备提供补贴。

推理：答案选项需指出“可以将原木转化为有潜在价值的产品”或“将原木转化为有潜在价值的产品真的可以产生销售收入”。

选项分析：

(A) 林业部在过去六年中将森林疏伐的预算增加了 15%。与伐木的预算无关。

(B) 如果能吸引一家大型跨国木材加工企业进入 X 国，将有助于降低失业率。与失业率无关。

(C) 林业部计划仔细确定有关原木的尺寸规格。与原木的尺寸规定无关。

(D) 从 X 国森林中取出的小树的木材类型并不都适合用作地板或木屑的原料。如果不适合用作原料，说明目标无法达成。此选项可以削弱。

(E) 正确。在 X 国，一座典型的建筑可以用价格合理的木屑来取暖，其费用仅为使用传统燃料费用的 20% 左右。此选项指出这些“有价值的产品”能卖出去，即价格不能比传统的产品更贵，这样可以真的达成目标。

8. The Internet service provider CableWeb is spending billions of dollars to expand its network to meet the growing demand for Internet service. CableWeb cannot raise subscription rates to pay for this expansion, or else many of its subscribers will switch to competing service providers. Thus, to generate funds for this expansion, CableWeb plans to require payments from owners of commercial Websites each time a CableWeb subscriber accesses those sites.

In order to evaluate how likely CableWeb's planned required payments will be to generate funds for expansion, it would be most helpful to know which of the following?

(A) To what extent owners of commercial Websites depend on revenues generated from consumers who access their sites via CableWeb's services

(B) How many Internet service providers competing with CableWeb are also expanding their networks to meet the growing demand for Internet service

(C) By what methods other than demanding payments from owners of commercial Websites CableWeb could feasibly pay to expand its network

(D) How CableWeb's subscription rates currently compare with those of its main competitors

(E) Whether any CableWeb subscribers would be willing to pay a fee each time they access a commercial Website

类别：方案

目标：为此次扩张创造资金。

方案：CableWeb 的方案要求商业网站的所有者在每次 CableWeb 用户访问这些网站时支付费用。

推理：答案选项需指出“商业网站是否真的会支付这笔费用”或“支付这笔费用能否真的获得资金”。

选项分析：

(A) 正确。商业网站的所有者在多大程度上依赖于由通过 CableWeb 服务访问其网站的消费者所产生的收入。如果这样的程度很低，那么收费方案很容易导致根本没人交钱，此方案从最开始就无法实施。

(B) 有多少与 CableWeb 竞争的互联网服务提供商也在扩大他们的网络，以满足对互联网服务日益增长的需求。与竞争对手是否在扩张无关。

(C) 除了要求商业网站的所有者付款外，CableWeb 还可以通过什么方法来支付扩大其网络的费用。与其他方法无关，除非是“更经济、更有效”的方法。

(D) 目前 CableWeb 的收费标准与它的主要竞争对手的收费标准相比如何。收费标准的对比与讨论无关。

(E) 是否有 CableWeb 的用户愿意在每次访问商业网站时支付费用。与用户是否愿意交钱无关，我们探讨的是“商业网站”交钱是否有利于达成目标。

9. Career counselor: Research has found that people who hunt extensively for the highest-paying careers earn, on average, about 20 percent more than other people. But they also experience less job satisfaction and more anxiety at work. This suggests that to find a fulfilling career, it is best not to focus primarily on maximizing your salary.

Which of the following, if true, would most weaken the career counselor's argument?

(A) People who find their careers fulfilling generally report higher job satisfaction and less anxiety at work than do people who find their careers unfulfilling.

(B) Most people who report that their careers are exceptionally fulfilling earn relatively modest salaries.

(C) On average, people who tend in general to feel less satisfaction and more anxiety are more likely to hunt extensively for the highest-paying careers.

(D) On average, people who hunt extensively for the most fulfilling careers, with little regard for salary, still earn more than other people.

(E) In some situations, it is much more important to find a career offering a higher salary than a career that is fulfilling.

类别：因果推理

前提：研究发现，广泛寻找高薪职业的人，平均比其他人多赚20%左右。但他们在工作中也会体验到较少的工作满意度和更多的焦虑感。

结论：为了找到一个令人满意的职业，最好不要主要关注使你的薪水最大化。

推理：

纯粹巧合（削弱）：指出“使薪水最大化”和“对工作不满意”之间不存在能产生因果关系的原理；或指出在其他场景中，“使薪水最大化”和“对工作不满意”不再具备正相关的关系。

因果倒置（削弱）：指出那些本来就对工作不满意的人才去追求薪水最大化。
他因导致结果（削弱）：给出其他可能导致对工作不满意的原因。

选项分析：

(A) 认为自己的职业有成就感的人，通常比认为自己的职业没有成就感的人，有更高的工作满意度和更少的工作焦虑。此选项没有提及工作满意度和“薪水”的问题。

(B) 大多数报告自己的职业特别有成就感的人薪水相对较低。此选项加强了结论。

(C) 正确。平均来说，那些总体上感到不太满意和比较焦虑的人，更有可能广泛地寻找收入最高的职业。此选项直接从“因果倒置”的角度削弱结论。

(D) 平均来说，那些广泛寻找最有成就感职业的人，几乎不考虑薪水，仍然比其他人挣得多。我们就是在讨论“薪水”和“对工作满意”之间的关系，不能“不考虑薪水”。

(E) 在某些情况下，找到一个薪水更高的职业比找到一个令人有成就感的职业要重要得多。与哪种工作更重要无关。

10. Journalist: Researchers surveyed 1, 000 people who had experienced personal catastrophes. They found that those who formulated new personal goals after such disasters achieved personal readjustment more effectively and more quickly than those who had not done so. Assuming that the study was done correctly, it follows that to bring about readjustment in such people most quickly and most effectively, counselors should mainly focus on encouraging their clients to formulate new personal goals.

Which of the following most accurately describes a flaw in the journalist's reasoning?

(A) The journalist relies on the use of emotively colored language, rather than on the presentation of data.

(B) The argument fails to consider that the readjustment even of those who have not experienced personal catastrophes might be expedited by the formulation of new personal goals.

(C) The argument fails to consider whether, for those who have experienced personal catastrophes in their lives, maintaining previous personal goals might be beneficial for readjustment.

(D) The journalist's generalization about what counselors should do relies on a study that, in its collection of data, did not include a sample of counselors.

(E) The argument fails to consider the possibility that formation of new personal goals is an early stage of readjustment rather than its cause.

类别：因果推理

前提：那些在灾难后制订了新的个人目标的人比那些没有这样做的人更有效、更迅速地进行了个人调整。

结论：为了使这些人最迅速、最有效地重新适应，顾问应该鼓励客户制订新的个人目标。

推理：

纯粹巧合（削弱）：指出“制订新的个人目标”和“更有效、更迅速地进行个人调整”之间不存在能产生因果关系的原理；或在其他场景中，“制订新的个人目标”和“更有效、更迅速地进行个人调整”不再具备正相关的关系。

因果倒置（削弱）：指出那些本来就更容易重新适应的人才会制订新的个人目标。

他因导致结果（削弱）：给出其他可能导致更有效、更迅速地进行个人调整的原因。

选项分析：

(A) 该记者依赖于使用带有感情色彩的语言，而不是数据的呈现。与记者表达的方式无关。

(B) 该论点没有考虑到，即使是那些没有经历过个人灾难的人，也可能通过制订新的个人目标来加快调整。与没有经历过灾难的人无关。

(C) 该论点没有考虑到，对于那些在生活中经历过个人灾难的人来说，保持以前的个人目标是否有利于重新调整。与以前的目标无关。

(D) 记者对咨询师应该做什么的归纳依赖于一项研究，而这项研究在收集数据时并没有包括咨询师的样本。与咨询师的情况无关。

(E) 正确。该论点没有考虑到，新的个人目标的形成是重新适应的早期阶段，而不是其原因的可能性。本选项直接指出了“因果倒置”这一谬误方向。

第三章

综合提高训练

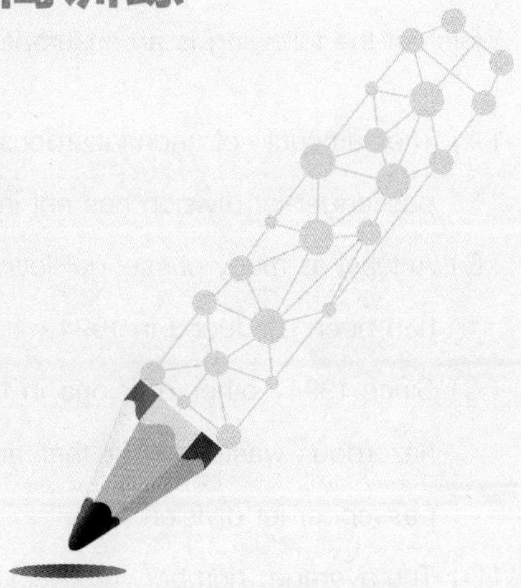

第一组

1. In response to mounting public concern, an airplane manufacturer implemented a program with the well-publicized goal of reducing by half the total yearly amount of hazardous waste generated by its passenger-jet division. When the program began in 1994, the division's hazardous waste output was 90 pounds per production worker; last year it was 40 pounds per production worker. Clearly, therefore, charges that the manufacturer's program has not met its goal are false.

Which of the following is an assumption on which the argument depends?

(A) The amount of nonhazardous waste generated each year by the passenger-jet division has not increased significantly since 1994.

(B) At least as many passenger jets were produced by the division last year as had been produced in 1994.

(C) Since 1994, other divisions in the company have achieved reductions in hazardous waste output that are at least equal to that achieved in the passenger-jet division.

(D) The average number of weekly hours per production worker in the passenger-jet division was not significantly greater last year than it was in 1994.

(E) The number of production workers assigned to the passenger-jet division was not significantly less in 1994 than it was last year.

2. City official: At City Hospital, uninsured patients tend to have shorter stays and fewer procedures performed than do insured patients, even though insured

patients, on average, have slightly less serious medical problems at the time of admission to the hospital than uninsured patients have. Critics of the hospital have concluded that **the uninsured patients are not receiving proper medical care**. However, **this conclusion is almost certainly false**. Careful investigation has recently shown two things: insured patients have much longer stays in the hospital than necessary, and they tend to have more procedures performed than are medically necessary.

In the city official's argument, the two boldface portions play which of the following roles?

(A) The first states the conclusion of the city official's argument; the second provides support for that conclusion.

(B) The first is used to support the conclusion of the city official's argument; the second states that conclusion.

(C) The first was used to support the conclusion drawn by hospital critics; the second states the position that the city official's argument opposes.

(D) The first was used to support the conclusion drawn by hospital critics; the second provides support for the conclusion of the city official's argument.

(E) The first states the position that the city official's argument opposes; the second states the conclusion of the city official's argument.

3. Last year, farmers who sold their corn early come to regret it as rising demand pushed corn prices higher later in the year. This year, farmers are experiencing record harvest, but demand for corn is expected to continue to rise, which suggests that prices will rise again this year. Nevertheless, most farmers are already seeking to sell large quantities of corn.

Which of the following, if true, provides the best explanation for the farmers' position?

(A) While this year's record harvest was unexpected, next years' harvest is predicted to equal or surpass it.

(B) Consumption of corn-based products has increased over the past decade and is likely to continue to do so.

(C) Corn stored in poorly produced facilities is likely to spoil or crack, rendering it useless for many applications.

(D) Most farmers do not have sufficient storage capacity available to them for the record amounts of corn from this year's harvest.

(E) Many manufacturers of corn-based products want to buy corn directly from nearby farmers to guarantee necessary quality.

4. Fact: Asthma, a bronchial condition, is much less common ailment than hay fever, an allergic inflammation of the nasal passages.

Fact: Over 95 percent of people who have asthma also suffer from hay fever.

If the information given as facts above is true, which of the following must also be true?

(A) Hay fever is a prerequisite for the development of asthma.

(B) Asthma is a prerequisite for the development of hay fever.

(C) Those who have neither hay fever nor asthma comprise less than 5 percent of the total population.

(D) The number of people who have both of these ailments is greater than the number of people who have only one of them.

(E) The percentage of people suffering from hay fever who also have asthma is lower than 95 percent.

5. Compared to regular automobile fuel, low-sulfur automobile fuel produces a lower volume of harmful emissions such as sulfur oxides. In an attempt to

improve the air quality in Litania, the Litanian government has cut the excise tax on low-sulfur fuel to a point well below that of regular fuel. Since automobiles perform equally well with either fuel, the government hopes that such a tax cut will encourage drivers to switch from regular fuel to the less polluting low-sulfur fuel.

Which of the following, if true, would most seriously call into question the Litanian government's plan to use this tax cut to improve the air quality?

(A) Automobile fuel is not the sole source of sulfur oxides in the atmosphere.

(B) In terms of sulfur oxide emissions, the air quality of Litania is not as bad as the air quality of neighboring countries.

(C) Some drivers in Litania are unaware of the availability of low-sulfur fuel as an alternative to regular fuel.

(D) The cut in the excise tax for low-sulfur fuel is large enough to encourage significantly greater automobile use.

(E) Many Litanian drivers remain unconvinced that automobile emissions are a significant threat to the air quality.

6. Psychologist: On average, people who read at least one book a month for pleasure go far more frequently to museums, concerts, and sporting events than do people who read less than one book a year. This shows that recreational reading tends to promote a healthy interest in other social and cultural activities.

The psychologist's argument is most vulnerable to criticism on which of the following grounds?

(A) It overlooks the possibility that the more time one spends on recreational reading, on average, the less time will remain available for other social and cultural activities.

(B) It takes for granted that the greater a person's level of interest in an activity, the more often the person participates in that activity.

(C) It overlooks the possibility that two or more phenomena may tend to occur together even if none of those phenomena causally contributes to any of the others

(D) It fails to adequately address the possibility that, even if one phenomenon tends to produce a certain effect, other phenomena may also contribute to that effect.

(E) It takes for granted that frequent attendance of museums, concerts, and sporting events does not tend to reduce interest in recreational reading.

7. Brochure: Help conserve our city's water supply. By converting the landscaping in your yard to a water-conserving landscape, you can greatly reduce your outdoor water use. A water-conserving landscape is natural and attractive, and it also saves you money.

Criticism: For most people with yards, the savings from converting to a water-conserving landscape cannot justify the expense of new landscaping, since typically the conversion would save less than twenty dollars on a homeowner's yearly water bills.

Which of the following, if true, provides the best basis for a rebuttal of the criticism?

(A) Even homeowners whose yards do not have water-conserving landscapes can conserve water by installing water-saving devices in their homes.

(B) A conventional landscape generally requires a much greater expenditure on fertilizer and herbicide than does a water-conserving landscape.

(C) A significant proportion of the residents of the city live in buildings that do not have yards.

(D) It costs no more to put in water-conserving landscaping than it does to put in conventional landscaping.
(E) Some homeowners use more water to maintain their yards than they use for all other purposes combined.

8. Unless tiger hunting decreases, tigers will soon be extinct in the wild. The countries in which the tigers' habitats are located are currently debating joint legislation that would ban tiger hunting. Thus, if these countries can successfully enforce this legislation, the survival of tigers in the wild will be ensured.

The reasoning in the argument is most vulnerable to criticism on the grounds that the argument

(A) assumes without sufficient warrant that a ban on tiger hunting could be successfully enforced
(B) considers the effects of hunting on tigers without also considering the effects of hunting on other endangered animal species
(C) fails to take into account how often tiger hunters are unsuccessful in their attempts to kill tigers
(D) neglects to consider the results of governmental attempts in the past to limit tiger hunting
(E) takes the removal of an impediment to the tigers' survival as a guarantee of their survival

9. Editorial: As societies grow richer, the service sector of the economy—education, health, banking, and so on—grows in importance relative to energy-intensive activities such as steel production. This shift lowers the ratio of carbon emissions to dollars of economic production. Hence, as societies grow richer, the totality of their carbon emissions declines.

The editorial's reasoning is most vulnerable to criticism on which of the following grounds?

(A) It confuses absolute decline with relative decline with respect to a growing quantity

(B) It takes for granted that growth in the services sector of a society's economy inevitably causes that society to grow richer.

(C) It overlooks the possibility that there may be an observable correlation between two phenomena even if neither phenomenon causally contributes to the other.

(D) It fails to adequately address the possibility that energy-intensive activities such as steel production tend to become more energy efficient as societies grow richer.

(E) It confuses a claim about carbon emissions with a claim about dollars of economic production.

10. During excavations for a highway, workers just uncovered a hoard of over 30,000 eighteenth-century gold coins. These coins, most in superb condition, are of a type that has long been quite rare and correspondingly highly valued. Several coins of the same date and type recently sold at auction for over $5,000 each, despite being in poor condition. Clearly, therefore, in today's market, this hoard, if sold, is bound to bring in at least $150,000,000.

Which of the following is an assumption on which the argument depends?

(A) The workers who uncovered the coins did not take for themselves a significant number of coins before reporting their find.

(B) The hoard of coins would not bring in more if broken down into small parcels to be sold individually than if sold as a single quantity.

(C) The appearance on the market of thousands of additional coins of this type will not significantly affect the status of these coins as rarities.

(D) It is not possible to determine whether the hoard of coins was buried by people who were then its legal owners.

(E) Whoever might sell the coins would be sufficiently knowledgeable about such sales to secure the highest possible price for the coins.

答案及解析

1. In response to mounting public concern, an airplane manufacturer implemented a program with the well-publicized goal of reducing by half the total yearly amount of hazardous waste generated by its passenger-jet division. When the program began in 1994, the division's hazardous waste output was 90 pounds per production worker; last year it was 40 pounds per production worker. Clearly, therefore, charges that the manufacturer's program has not met its goal are false.

Which of the following is an assumption on which the argument depends?

(A) The amount of nonhazardous waste generated each year by the passenger-jet division has not increased significantly since 1994.

(B) At least as many passenger jets were produced by the division last year as had been produced in 1994.

(C) Since 1994, other divisions in the company have achieved reductions in hazardous waste output that are at least equal to that achieved in the passenger-jet division.

(D) The average number of weekly hours per production worker in the passenger-jet division was not significantly greater last year than it was in 1994.

(E) The number of production workers assigned to the passenger-jet division was not significantly less in 1994 than it was last year.

类别：普通预测推理

前提：当项目开始于 1994 年的时候，有害污染物的人均排放量是 90 磅/人；去年是 40 磅/人。

结论：目标没达到的指控是错误的（目标达到 = 有害污染物的排放总量减少了一半）。

推理：结论“目标没达到的指控是错误的”的充分条件应为“人均×人数的结果减少一半”。前提只讲了人均，忽略了人数的情况。

选项分析：

(A) 自 1994 年以来，飞机制造商客机部门无害污染物的排放量并没有显著增加。与“无害污染物的排放量”无关。

(B) 去年制造的客机数量至少和 1994 年制造的客机数量一样。与客机制造量无关。

(C) 自 1994 年以来，其他部门有害污染物排放量的减少至少和客机制造部门的一样。与其他部门无关。

(D) 去年每个员工的平均工作时间没有明显长于 1994 年每个员工的平均工作时间。与员工工作时长无关。

(E) 正确。1994 年客机制造部门的员工数没有明显少于去年的员工数。只有当员工数量几乎稳定的时候，人均排放减少量才会等于总体排放减少量。

2. City official: At City Hospital, uninsured patients tend to have shorter stays and fewer procedures performed than do insured patients, even though insured patients, on average, have slightly less serious medical problems at the time of admission to the hospital than uninsured patients have. Critics of the hospital have concluded that **the uninsured patients are not receiving proper medical care**. However, **this conclusion is almost certainly false**. Careful investigation has recently shown two things: insured patients have much longer stays in the hospital than necessary, and they tend to have more procedures performed than are medically necessary.

In the city official's argument, the two boldface portions play which of the following roles?

(A) The first states the conclusion of the city official's argument; the second provides support for that conclusion.

(B) The first is used to support the conclusion of the city official's argument; the second states that conclusion.

(C) The first was used to support the conclusion drawn by hospital critics; the second states the position that the city official's argument opposes.

(D) The first was used to support the conclusion drawn by hospital critics; the second provides support for the conclusion of the city official's argument.

(E) The first states the position that the city official's argument opposes; the second states the conclusion of the city official's argument.

类别：分析论证

推理：阅读文段，可以确定主结论为：This conclusion is almost certainly false.（第二个黑体句）第一个黑体句是主结论反对的一个观点。

选项分析：

(A) 第一个黑体句指出了市政府官员论证的结论；第二个黑体句为该结论提供支持。

(B) 第一个黑体句是用来支持该市政府官员论证的结论；第二个黑体句陈述了该结论。

(C) 第一个黑体句被用来支持医院批评者得出的结论；第二个黑体句陈述了市政府官员的论证所反对的立场。

(D) 第一个黑体句用来支持医院批评者得出的结论；第二个黑体句为市政府官员的论证结论提供支持。

(E) 正确。第一个黑体句陈述了市政府官员的论证所反对的立场；第二个黑体句陈述了市政府官员论证的结论。

3. Last year, farmers who sold their corn early come to regret it as rising demand pushed corn prices higher later in the year. This year, farmers are experiencing record harvest, but demand for corn is expected to continue to rise, which suggests that prices will rise again this year. Nevertheless, most farmers are already seeking to sell large quantities of corn.

Which of the following, if true, provides the best explanation for the farmers' position?

(A) While this year's record harvest was unexpected, next years' harvest is predicted to equal or surpass it.

(B) Consumption of corn-based products has increased over the past decade and is likely to continue to do so.

(C) Corn stored in poorly produced facilities is likely to spoil or crack, rendering it useless for many applications.

(D) Most farmers do not have sufficient storage capacity available to them for the record amounts of corn from this year's harvest.

(E) Many manufacturers of corn-based products want to buy corn directly from nearby farmers to guarantee necessary quality.

类别：构建论证中的“现象解释”。

推理：文中需要解释的矛盾点是：玉米以后会涨价，但很多农民已经在设法出售大量玉米。

选项分析：

(A) 虽然今年创纪录的收成是出乎意料的，但预计明年的收成将与之持平或超过。与明年的玉米收成无关。

(B) 在过去的十年里，玉米类产品的消费量已经增加，并可能继续增加。无法解释农民为什么着急卖掉玉米。

(C) 储存在生产条件差的设施中的玉米很可能会变质或开裂，使其在许多应用中失去作用。文中并没有提到农民现在储存玉米的设施情况如何。如果选项或者文段补充一句“农民现在储存玉米的设施很差”，那么可以考虑此选项。

(D) 正确。大多数农民没有足够的储存能力来储存今年收获的创纪录数量的玉米。解释了农民为什么着急卖玉米，因为没有地方储存了。

(E) 许多玉米类产品的制造商希望直接从附近的农民那里购买玉米，以保证必要的质量。即使制造商希望直接从农民那里购买玉米，也无法解释为什么农民不等玉米涨价之后再卖。

4. Fact：Asthma，a bronchial condition，is much less common ailment than hay fever，an allergic inflammation of the nasal passages.

Fact：Over 95 percent of people who have asthma also suffer from hay fever.

If the information given as facts above is true，which of the following must also be true?

(A) Hay fever is a prerequisite for the development of asthma.
(B) Asthma is a prerequisite for the development of hay fever.
(C) Those who have neither hay fever nor asthma comprise less than 5 percent of the total population.
(D) The number of people who have both of these ailments is greater than the number of people who have only one of them.
(E) The percentage of people suffering from hay fever who also have asthma is lower than 95 percent.

类别：构建论证中的“确定结论”

推理：

原文有两个事实。

事实 1：哮喘是一种支气管疾病，比花粉热（一种鼻腔过敏性炎症）要少得多。

事实 2：超过 95% 的哮喘患者也患有花粉热。

本题比较像数学题，可以用韦恩图来帮助理解。asthma 是小圆，hay fever 是大圆，一个小圆和一个大圆交叉，交叉面积占小圆的 95%，那么交叉面积占大圆的比例肯定小于大圆的 95%。

选项分析：

(A) 花粉热是哮喘发展的先决条件。原文没有提谁是谁的先决条件。

(B) 哮喘是花粉热发展的先决条件。原文没有提谁是谁的先决条件。

(C) 既没有花粉热也没有哮喘的人占总人口的不到5%。无法确定均不患两种疾病的人数占比。

(D) 同时患有这两种疾病的人数多于只患有其中一种疾病的人数。我们无法确定交叉面积占大圆的多少。

(E) 正确。花粉热患者同时患有哮喘的比例低于95%。

5. Compared to regular automobile fuel, low-sulfur automobile fuel produces a lower volume of harmful emissions such as sulfur oxides. In an attempt to improve the air quality in Litania, the Litanian government has cut the excise tax on low-sulfur fuel to a point well below that of regular fuel. Since automobiles perform equally well with either fuel, the government hopes that such a tax cut will encourage drivers to switch from regular fuel to the less polluting low-sulfur fuel.

Which of the following, if true, would most seriously call into question the Litanian government's plan to use this tax cut to improve the air quality?

(A) Automobile fuel is not the sole source of sulfur oxides in the atmosphere.

(B) In terms of sulfur oxide emissions, the air quality of Litania is not as bad as the air quality of neighboring countries.

(C) Some drivers in Litania are unaware of the availability of low-sulfur fuel as an alternative to regular fuel.

(D) The cut in the excise tax for low-sulfur fuel is large enough to encourage significantly greater automobile use.

(E) Many Litanian drivers remain unconvinced that automobile emissions are a significant threat to the air quality.

类别：方案

目标：改善利塔尼亚的空气质量。

方案：利塔尼亚政府已经将低硫燃料的消费税削减到远低于普通燃料的水平。

推理：答案选项需指出“无法将低硫燃料的消费税削减到远低于普通燃料的水平”或“这么做无法真的改善空气质量”。

选项分析：

(A) 汽车燃料不是大气中硫氧化物的唯一来源。无论有无其他来源，汽车如果能改用低硫燃料就可以改善空气质量。

(B) 就硫氧化物的排放而言，利塔尼亚的空气质量没有邻国的空气质量那么差。与比较邻国的空气质量无关。

(C) 利塔尼亚的一些司机不知道低硫燃料可以替代普通燃料。与知不知道无关，知道了也不一定会真的去做。

(D) 正确。对低硫燃料消费税的削减，幅度大到足以鼓励更多地使用汽车。如果汽车的使用提高，那么自然该政策虽然会导致现有汽车减排，但更多地使用汽车可能会增加总排放。

(E) 许多利坦尼亚的司机仍然不相信汽车排放是对空气质量的一个重大威胁。与司机是否相信无关。

6. Psychologist: On average, people who read at least one book a month for pleasure go far more frequently to museums, concerts, and sporting events than do people who read less than one book a year. This shows that recreational reading tends to promote a healthy interest in other social and cultural activities.

The psychologist's argument is most vulnerable to criticism on which of the following grounds?

(A) It overlooks the possibility that the more time one spends on recreational reading, on average, the less time will remain available for other social and cultural activities.

(B) It takes for granted that the greater a person's level of interest in an activity, the more often the person participates in that activity.

(C) It overlooks the possibility that two or more phenomena may tend to occur together even if none of those phenomena causally contributes to any of the others

(D) It fails to adequately address the possibility that, even if one phenomenon tends to produce a certain effect, other phenomena may also contribute to that effect.

(E) It takes for granted that frequent attendance of museums, concerts, and sporting events does not tend to reduce interest in recreational reading.

类别：因果推理

前提：平均来说，每月至少读一本书消遣的人去博物馆、音乐会和体育赛事的频率要远远高于每年读不到一本书的人。

结论：休闲阅读往往会促进人们对其他社会和文化活动产生对身体有益的兴趣。

推理：

纯粹巧合：在其他场景下，“休闲阅读”和“人们对其他社会和文化活动产生对身体有益的兴趣”没有同时存在；不存在“休闲阅读”导致“人们对其他社会和文化活动产生对身体有益的兴趣”的原理。

他因导致结果：存在其他导致“人们对其他社会和文化活动产生对身体有益的兴趣”的原因。

因果倒置：有可能是“人们对其他社会和文化活动产生对身体有益的兴趣”导致“休闲阅读”。

选项分析：

(A) 它忽略了这样一种可能性，即一个人在休闲阅读上花的时间越多，平均来说，可用于其他社会和文化活动的时间就越少。

(B) 它想当然地认为，一个人对某项活动的兴趣程度越高，这个人就越经常参与这项活动。

(C) 正确。它忽略了两种或更多种现象往往同时发生的可能性，即使这些现象中没有任何一种对其他现象有因果关系。本选项直接指出了“纯粹巧合”这一谬误的定义。

(D) 它没有充分解决这样一种可能性，即使一种现象倾向于产生某种效果，其他现象也可能对这种效果起作用。本选项是比较明显的干扰项，因为它很像是在给出“他因导致结果”这一方向。但是，他因要求的是这个原因必然和结论中的原因是互斥的，不能同时存在。如果两个原因本身可以同时存在，那么原结论中的因果关系就依然成立。因此，本选项是错误的，因为它指出的是两个可以同时存在的原因，不属于“他因导致结果”这一谬误方向。

(E) 它想当然地认为，经常去博物馆、音乐会和体育赛事并不会使人们降低对休闲阅读的兴趣。

7. Brochure: Help conserve our city's water supply. By converting the landscaping in your yard to a water-conserving landscape, you can greatly reduce your outdoor water use. A water-conserving landscape is natural and attractive, and it also saves you money.

Criticism: For most people with yards, the savings from converting to a water-conserving landscape cannot justify the expense of new landscaping, since typically the conversion would save less than twenty dollars on a homeowner's yearly water bills.

Which of the following, if true, provides the best basis for a rebuttal of the criticism?

(A) Even homeowners whose yards do not have water-conserving landscapes can conserve water by installing water-saving devices in their homes.

(B) A conventional landscape generally requires a much greater expenditure on fertilizer and herbicide than does a water-conserving landscape.

(C) A significant proportion of the residents of the city live in buildings that do not have yards.

(D) It costs no more to put in water-conserving landscaping than it does to put in conventional landscaping.

(E) Some homeowners use more water to maintain their yards than they use for all other purposes combined.

类别：普通预测推理

前提：节水景观节省的水费不到 20 美元。

结论：节水景观节省的钱没超过花费。

推理：结论“节水景观节省的钱没超过花费”的充分条件应为“节水景观节省的所有钱没超过其所有的花费”，但前提只提了节省的 20 美元水费。

选项分析：

(A) 就算那些院子里没有安装节水景观的房屋所有者也可以通过在房屋里安装节水设备来节水。本选项没有提及节水景观能省下的钱。

(B) 正确。传统景观相对于节水景观来说，需要在肥料和杀虫剂上花费更多。本选项给出了节水景观能省钱的另外一个方面，即不单单是水费可以节省钱，还可以节省其他方面的开支。

(C) 很多人住在没有院子的房子里。选项没有提及节水景观能省下的钱。

(D) 安装节水景观不会比安装传统景观更贵。本选项对比的是传统景观和节水景观的价格，并不是节水景观在何处可以省钱。

(E) 有些房屋所有者在维护院子上用的水比他在所有其他目的上用的水都多。选项没有提及节水景观能省下的钱。

8. Unless tiger hunting decreases, tigers will soon be extinct in the wild. The countries in which the tigers' habitats are located are currently debating joint legislation that would ban tiger hunting. Thus, if these countries can

successfully enforce this legislation, the survival of tigers in the wild will be ensured.

The reasoning in the argument is most vulnerable to criticism on the grounds that the argument

(A) assumes without sufficient warrant that a ban on tiger hunting could be successfully enforced

(B) considers the effects of hunting on tigers without also considering the effects of hunting on other endangered animal species

(C) fails to take into account how often tiger hunters are unsuccessful in their attempts to kill tigers

(D) neglects to consider the results of governmental attempts in the past to limit tiger hunting

(E) takes the removal of an impediment to the tigers' survival as a guarantee of their survival

类别：普通预测推理

前提：这些国家可以成功立法（禁止捕猎老虎）。

结论：老虎在野外的生存得到保证。

推理：结论“老虎在野外的生存得到保证”的充分条件应为“国家不仅立法且可以严格执行，堵住所有会威胁老虎在野外生存的渠道”。

选项分析：

(A) 假设如果没有足够的依据，限制捕猎老虎的法令可以成功实施。

(B) 考虑捕猎老虎带来的影响，而没有考虑捕猎其他濒危动物所带来的影响。

(C) 没有考虑猎人捕杀老虎失败的概率有多少。

(D) 忽略了考虑政府过去试图限制捕猎老虎的结果。

(E) 正确。认为消除某一个妨碍老虎生存的理由就可以保证老虎的生存。此选项指出禁止捕猎老虎并不是保证老虎生存的唯一因素，还需要考虑其他情况。

9. Editorial: As societies grow richer, the service sector of the economy—education, health, banking, and so on—grows in importance relative to energy-intensive activities such as steel production. This shift lowers the ratio of carbon emissions to dollars of economic production. Hence, as societies grow richer, the totality of their carbon emissions declines.

The editorial's reasoning is most vulnerable to criticism on which of the following grounds?

(A) It confuses absolute decline with relative decline with respect to a growing quantity

(B) It takes for granted that growth in the services sector of a society's economy inevitably causes that society to grow richer.

(C) It overlooks the possibility that there may be an observable correlation between two phenomena even if neither phenomenon causally contributes to the other.

(D) It fails to adequately address the possibility that energy-intensive activities such as steel production tend to become more energy efficient as societies grow richer.

(E) It confuses a claim about carbon emissions with a claim about dollars of economic production.

类别：普通预测推理

前提：从第二到第三产业这一转变降低了碳排放量与经济生产总量的比值。

结论：随着社会越来越富裕，其碳排放总量也在下降。

推理：结论“碳排放总量下降”的充分条件应为“碳排放量与经济生产总量的比值×经济生产总量”的值下降。而前提提到的是“碳排放量:经济总值”，结论提到的是“碳排放量”，很明显，两者不可以直接画等号。

选项分析：

(A) 正确。它混淆了绝对下降和有关增长量的相对下降。本选项直接指出了前提的偏差点。

(B) 它想当然地认为社会经济中服务部门的增长不可避免地导致该社会变得更富有。

(C) 它忽略了两种现象之间可能存在可观察到的相关性，即使两种现象都没有对另一种现象产生因果关系。

(D) 它没有充分解决能源密集型活动，如钢铁生产往往随着社会的富裕而变得更加节能。

(E) 它混淆了关于碳排放的说法和关于经济产值是多少美元的说法。

10. During excavations for a highway, workers just uncovered a hoard of over 30, 000 eighteenth-century gold coins. These coins, most in superb condition, are of a type that has long been quite rare and correspondingly highly valued. Several coins of the same date and type recently sold at auction for over $5, 000 each, despite being in poor condition. Clearly, therefore, in today's market, this hoard, if sold, is bound to bring in at least $150, 000, 000.

Which of the following is an assumption on which the argument depends?

(A) The workers who uncovered the coins did not take for themselves a significant number of coins before reporting their find.

(B) The hoard of coins would not bring in more if broken down into small parcels to be sold individually than if sold as a single quantity.

(C) The appearance on the market of thousands of additional coins of this type will not significantly affect the status of these coins as rarities.

(D) It is not possible to determine whether the hoard of coins was buried by people who were then its legal owners.

(E) Whoever might sell the coins would be sufficiently knowledgeable about such sales to secure the highest possible price for the coins.

类别：普通预测推理

前提：最近有几枚日期和类型相同的硬币在拍卖会上以每枚超过5000美元的价格售出，尽管它们的状况不佳。

结论：在今天的市场上，这批硬币如果被卖掉，势必会带来至少150万美元的收入。

推理：结论“这批硬币会带来至少150万美元的收入”的充分条件应为“这批硬币的每一枚依然能卖到5000美元以上”。但前提讲的是某几枚硬币在拍卖会上能卖出5000美元。因此，我们需要确保新得到的硬币能像之前的硬币一样，价值不会有很大变化。

选项分析：

(A) 挖掘硬币的工人在报告他们的发现之前，并没有将大量的硬币据为己有。即使在报告前据为己有，能确保上报的硬币有3万枚即可。

(B) 这批硬币如果被分解成小包单独出售，不会比整体出售带来更多的收益。与卖出的形式无关。

(C) 正确。市场上出现数以千计的这种类型的硬币，不会对这些硬币的稀有地位产生重大影响。本选项很直白地修正了偏差。

(D) 不可能确定这批硬币是否是由当时的合法所有者埋藏的。由谁埋葬与讨论无关。

(E) 无论可能出售这些硬币的人是谁，都会对这种销售有足够的了解，以确保这些硬币的销售价格最高。我们只需确保这批硬币“至少”能卖150万美元即可，与“最高”能卖多少美元无关。

第二组

1. Real estate agent: Over the past decade, our region's population has increased over 10 percent, and so have inflation-adjusted sales prices of local homes. Demographers project that the population will rise even more rapidly over the coming decade. Population growth increases demand for homes, and home prices rise when demand rises relative to supply. Thus, housing prices over the next decade will probably also keep rising at least as rapidly as they have been.

Which of the following is an assumption the real estate agent's argument requires?

(A) Over the coming decade, the supply of homes in the region will not increase at a significantly greater rate than the demand for homes there.

(B) Other things being equal, rising home prices tend to slow regional population growth.

(C) Demographers have projected that the regional population growth will cause a proportional increase in inflation-adjusted sales prices of local homes.

(D) Even if the regional population does not rise over the coming decade, local housing prices will probably continue to rise.

(E) Rising prices for a product during one decade usually, if not always, indicate that prices for that product will continue to rise during the next decade.

2. Economist: Tropicorp, which constantly seeks profitable investment opportunities, has been buying and clearing sections of tropical forest for cattle ranching, although pastures newly created there become useless for grazing after just a

few years. The company has not gone into rubber tapping, even though **greater profits can be made from rubber tapping**, which leaves the forest intact. Thus, some environmentalists conclude that **Tropicorp has not acted wholly out of economic self-interest**. However, these environmentalists are probably wrong. The initial investment required for a successful rubber-tapping operation is larger than that needed for a cattle ranch. Furthermore, there is a shortage of workers employable in rubber-tapping operations, and finally, taxes are higher on profits from rubber tapping than on profits from cattle ranching.

In the economist's argument, the two boldfaced portions play which of the following roles?

(A) The first supports the conclusion of the economist's argument; the second calls that conclusion into question.

(B) The first states the conclusion of the economist's argument; the second supports that conclusion.

(C) The first supports the environmentalists' conclusion; the second states that conclusion.

(D) The first states the environmentalists' conclusion; the second states the conclusion of the economist's argument.

(E) Each supports the conclusion of the economist's argument.

3. In a national survey, 50 percent of the surveyed households reported having credit card debt. But credit card companies reported that 76 percent of the nation's households owed them money. However, detailed follow-up investigations found that almost none of the surveyed households had reported their credit card debt inaccurately and that the credit card companies' reports were also accurate.

Which of the following, if true, would most help resolve the apparent discrepancy described above?

(A) Many of the surveyed households did not answer the question about their credit card debt.

(B) In several other surveys, even households that tried to report their debt levels honestly have tended to significantly underestimate their overall debt.

(C) In the nation as a whole, 50 percent of the households owe about 76 percent of the total household debt owed to credit card companies.

(D) The method of choosing households for the survey was unintentionally biased toward selecting households that were especially likely to have credit card debt.

(E) At least 26 percent of the surveyed households owed money to lenders other than credit card companies.

4. Which of the following most logically completes the argument below?

According to promotional material published by the city of Springfield, more tourists stay in hotels in Springfield than stay in the neighboring city of Harristown. A brochure from the largest hotel in Harristown claims that more tourists stay in that hotel than stay in the Royal Arms Hotel in Springfield. If both of these sources are accurate, however, the county's "Report on Tourism" must be in error in indicating that _______.

(A) more tourists stay in hotel accommodations in Harristown than stay in the Royal Arms Hotel

(B) the Royal Arms Hotel is the only hotel in Springfield

(C) there are several hotels in Harristown that are larger than the Royal Arms Hotel

(D) some of the tourists who have stayed in hotels in Harristown have also stayed in the Royal Arms Hotel

(E) some hotels in Harristown have fewer tourist guests each year than the Royal Arms Hotel has

5. Most cable television companies currently require customers to subscribe to packages of channels, but consumer groups have recently proposed legislation that would force the companies to offer a la carte pricing. Subscribers would pay less, argue the consumer groups, because they could purchase only the desired channels. However, the cable industry argues that under the current package pricing, popular channels subsidize less-popular ones, providing more options for viewers. For this reason, the industry claims that it is always cheaper for the consumer to purchase many bundled channels than to buy them individually.

Which of the following would be most important for the government to determine before deciding whether to require cable television companies to offer a la carte pricing in order to reduce consumer costs?

(A) Whether the total number of channels offered to consumers would decrease, along with programming diversity, as a result of the a la carte pricing structure

(B) Whether advertising revenue for the cable television companies would decrease as a result of the a la carte pricing structure

(C) Whether the vast majority of consumers would greatly reduce the number of channels purchased if given the option of purchasing them individually

(D) Whether cable and satellite companies currently have the ability to buy channels individually from programmers and content providers

(E) Whether a la carte subscribers would be required to have new television set-top boxes

6. New water-current patterns have brought more mixing of the Thalian Sea's shallow waters with nutrient-rich deeper waters. For coral and algae, both of which inhabit the shallow waters, growth rates depend principally on availability of nutrients and sufficient sunlight. The algae's rate of growth has increased, yet

the coral's growth has slowed. Since algae can block sunlight from reaching coral below, the slowed coral growth is probably attributable to increased algae growth.

Which of the following, if true, most strengthens the argument?

(A) The new water-current patterns in the Thalian Sea have increased the concentrations of industrial pollutants in its shallower parts.

(B) In those areas of the Thalian Sea where there has been the greatest increase in algae growth, the growth of coral has stopped altogether.

(C) In some areas of the Thalian Sea where shallow water receives abundant sunlight, there is no coral and there are only small amounts of algae.

(D) Many of the nutrients essential to the growth of algae play no role in the growth of coral.

(E) The mixing of the sea's shallow waters with nutrient-rich deeper waters is expected to diminish after this year.

7. Healthy lungs produce a natural antibiotic that protects them from infection by routinely killing harmful bacteria on airway surfaces. People with cystic fibrosis, however, are unable to fight off such bacteria, even though their lungs produce normal amounts of the antibiotic. The fluid on airway surfaces in the lungs of people with cystic fibrosis has an abnormally high salt concentration; accordingly, scientists hypothesize that the high salt concentration is what makes the antibiotic ineffective.

Which of the following, if true, most strongly supports the scientists' hypothesis?

(A) When the salt concentration of the fluid on the airway surfaces of healthy people is raised artificially, the salt concentration soon returns to normal.

(B) A sample of the antibiotic was capable of killing bacteria in an environment with an unusually low concentration of salt.

(C) When lung tissue from people with cystic fibrosis is maintained in a solution with a normal salt concentration, the tissue can resist bacteria.

(D) Many lung infections can be treated by applying synthetic antibiotics to the airway surfaces.

(E) High salt concentrations have an antibiotic effect in many circumstances.

8. Brant Lumber employs numerous workers to saw logs into boards. Brant is considering refitting its sawmill with computerized log-cutting machines that would saw the logs automatically and that would increase the number of logs the sawmill could process per day. The refit would allow Brant to reduce its workforce, but it would not increase the sawmill's output, since the sawmill already processes as many logs per day as Brant's timber sources can supply.

Which of the following, if true, casts the most doubt on the analysis of the effects of the refit on the sawmill's output?

(A) The computerized log-cutting machines cost far more to purchase than the machines currently installed in the sawmill did when they were new.

(B) Replacing some of the workers at the sawmill with computerized log-cutting machines would enable Brant to increase profits, even though those machines would not be used to capacity.

(C) By relocating the sawmill instead of refitting it, Brant would be able to increase the supply of lops available for the sawmill to process into boards.

(D) Less of a log is wasted when it is cut into boards by computerized log-cutting machines than when it is cut into boards by workers using the sawmill's current equipment.

(E) The computerized log-cutting machines can be adjusted far more quickly to cut boards of different dimensions than can the sawmill's current equipment.

9. Consultant: Innovation is fostered by frequent exchange and discussion of relevant ideas, but business work spaces divided into individual offices and

cubicles encourage employees to work in relative isolation. So, employees whose work requires them to innovate will generally be more productive if assigned to work at desks in large office space without dividing barriers.

Which of the following is an assumption on which the consultant's argument depends?

(A) In larger office spaces without dividing barriers, employees whose work requires them to innovate would more often discuss relevant ideas with each other than in other office settings.

(B) Discussing ideas at work reduces the time that employees would otherwise spend alone working on innovative ideas.

(C) In work spaces divided into individual offices and cubicles, employees whose work does not require them to innovate are sometimes more productive than in large open work spaces.

(D) Most of the ideas employees exchange and discuss at work are consciously considered by those employees to be directly relevant to innovation in their work.

(E) Business consultants generally recommend that innovative employees be assigned to work at desks in larger office spaces without dividing barriers if that will make them more productive.

10. In January of last year the Moviemania chain of movie theaters started propping its popcorn in canola oil, instead of the less healthful coconut oil that it had been using until then. Now Moviemania is planning to switch back, saying that the change has hurt popcorn sales. That claim is false, however, since according to Moviemania's own sales figures, Moviemania sold 5 percent more popcorn last year than in the previous year.

Which of the following, if true, most strongly supports the argument against Moviemania's claim?

(A) Total sales of all refreshments at Moviemania's movie theaters increased by less than 5 percent last year.

(B) Moviemania makes more money on food and beverages sold at its theaters than it does on sales of movie tickets.

(C) Moviemania's customers prefer the taste of popcorn popped in coconut oil to that of popcorn popped in canola oil.

(D) Total attendance at Moviemania's movie theaters was more than 20 percent higher last year than the year before.

(E) The year before last, Moviemania experienced a 10 percent increase in popcorn sales over the previous year.

答案及解析

1. Real estate agent：Over the past decade，our region's population has increased over 10 percent，and so have inflation-adjusted sales prices of local homes. Demographers project that the population will rise even more rapidly over the coming decade. Population growth increases demand for homes，and home prices rise when demand rises relative to supply. Thus，housing prices over the next decade will probably also keep rising at least as rapidly as they have been.

Which of the following is an assumption the real estate agent's argument requires？

(A) Over the coming decade，the supply of homes in the region will not increase at a significantly greater rate than the demand for homes there.
(B) Other things being equal，rising home prices tend to slow regional population growth.
(C) Demographers have projected that the regional population growth will cause a proportional increase in inflation-adjusted sales prices of local homes.
(D) Even if the regional population does not rise over the coming decade，local housing prices will probably continue to rise.
(E) Rising prices for a product during one decade usually，if not always，indicate that prices for that product will continue to rise during the next decade.

类别：普通预测推理

前提：人口增长增加了对住房的需求，而当需求相对于供应增加时，房屋价格就会上升。未来十年内人口会快速增长。

结论：未来十年内的住房价格可能也会持续上涨，至少和以前一样快。

推理，结论“未来十年内的住房价格上涨”的充分条件应为“需求相对于供应增加”。根据前提可知，房价受需求和供给的共同影响，且人口会影响需求。未来十年内人口会增加，就等于未来十年内对住房的需求会增加。所以若想要得出结论中的“房价会上涨”，我们需要考虑供给的情况。

选项分析：

(A) 正确。在未来的十年内，该地区的住房供应量不会以明显大于住房需求的速度增长。

(B) 在其他条件相同的情况下，住房价格的上涨往往会减缓该地区人口的增长。

(C) 人口学家预测，该地区人口增长将导致当地住房经通胀调整后的销售价格按比例上升。

(D) 即使该地区的人口在未来十年内不会增长，当地的住房价格也可能继续上涨。

(E) 一种产品的价格在十年内上涨，如果不总是如此的话，通常表明该产品的价格在下一个十年内将继续上涨。

2. Economist: Tropicorp, which constantly seeks profitable investment opportunities, has been buying and clearing sections of tropical forest for cattle ranching, although pastures newly created there become useless for grazing after just a few years. The company has not gone into rubber tapping, even though **greater profits can be made from rubber tapping**, which leaves the forest intact. Thus, some environmentalists conclude that **Tropicorp has not acted wholly out of economic self-interest**. However, these environmentalists are probably wrong. The initial investment required for a successful rubber-tapping operation is larger than that needed for a cattle ranch. Furthermore, there is a shortage of workers employable in rubber-tapping operations, and finally, taxes are higher on profits from rubber tapping than on profits from cattle ranching.

In the economist's argument, the two boldfaced portions play which of the following roles?

(A) The first supports the conclusion of the economist's argument; the second calls that conclusion into question.

(B) The first states the conclusion of the economist's argument; the second supports that conclusion.

(C) The first supports the environmentalists' conclusion; the second states that conclusion.

(D) The first states the environmentalists' conclusion; the second states the conclusion of the economist's argument.

(E) Each supports the conclusion of the economist's argument.

类别：分析论证

推理：阅读文段，可以确定主结论为：These environmentalists are probably wrong. 第一个黑体句是支持环境学家的结论的事实，第二个黑体句是环境学家的结论，亦是和主结论相反的结论。

选项分析：

(A) 第一个黑体句支持经济学家的论证结论；第二个黑体句对该结论提出质疑。

(B) 第一个黑体句指出了经济学家的论证结论；第二个黑体句支持该结论。

(C) 正确。第一个黑体句支持环保主义者的结论；第二个黑体句指出该结论。

(D) 第一个黑体句说的是环保主义者的结论；第二个黑体句说的是经济学家的论证结论。

(E) 两个黑体句均支持了经济学家的论证结论。

3. In a national survey, 50 percent of the surveyed households reported having credit card debt. But credit card companies reported that 76 percent of the nation's households owed them money. However, detailed follow-up

investigations found that almost none of the surveyed households had reported their credit card debt inaccurately and that the credit card companies' reports were also accurate.

Which of the following, if true, would most help resolve the apparent discrepancy described above?

(A) Many of the surveyed households did not answer the question about their credit card debt.

(B) In several other surveys, even households that tried to report their debt levels honestly have tended to significantly underestimate their overall debt.

(C) In the nation as a whole, 50 percent of the households owe about 76 percent of the total household debt owed to credit card companies.

(D) The method of choosing households for the survey was unintentionally biased toward selecting households that were especially likely to have credit card debt.

(E) At least 26 percent of the surveyed households owed money to lenders other than credit card companies.

类别：构建论证中的“现象解释”。

推理：阅读文段，文中需要解释的矛盾点是：50%的被调查家庭表示有信用卡债务，但信用卡公司称，76%的家庭欠他们的钱。

选项分析：

(A) 正确。许多被调查的家庭没有回答关于他们信用卡债务的问题。解释了原文中的偏差：50%的被调查家庭并不是全部。

(B) 在其他几项调查中，即使那些试图诚实报告他们的债务水平的家庭，也往往大大低估了他们的总体债务。

(C) 在全国范围内，50% 的家庭所欠的债务约占所有家庭欠信用卡公司债务总额的 76%。讨论的是家庭数量占比，与债务金额无关。

(D) 调查中选择家庭的方法无意中偏向于选择那些特别有可能有信用卡债务的家庭。

(E) 至少有 26% 的被调查家庭欠信用卡公司以外的放款人的钱。原文讨论的范围局限于“信用卡公司”，与其他债主无关。

4. Which of the following most logically completes the argument below?

According to promotional material published by the city of Springfield, more tourists stay in hotels in Springfield than stay in the neighboring city of Harristown. A brochure from the largest hotel in Harristown claims that more tourists stay in that hotel than stay in the Royal Arms Hotel in Springfield. If both of these sources are accurate, however, the county's "Report on Tourism" must be in error in indicating that ________.

(A) more tourists stay in hotel accommodations in Harristown than stay in the Royal Arms Hotel

(B) the Royal Arms Hotel is the only hotel in Springfield

(C) there are several hotels in Harristown that are larger than the Royal Arms Hotel

(D) some of the tourists who have stayed in hotels in Harristown have also stayed in the Royal Arms Hotel

(E) some hotels in Harristown have fewer tourist guests each year than the Royal Arms Hotel has

类别：构建论证中的“确定结论”

推理：

阅读文段，可以总结出两条信息：

(1) S 城的游客数量 >H 城的游客数量；

(2) S 城 RA 酒店的游客数量 < H 城最大酒店的游客数量。

注意文段最后让我们找出哪个选项一定“错误”。

选项分析：

(A) 住在 H 城的酒店的游客数量要多于住在 RA 酒店的游客数量。H 城最大酒店的游客数量已经比 RA 多了，那么 H 城的游客数量肯定比 RA 酒店的游客数量多。

(B) 正确。RA 酒店是 S 城唯一的酒店。如果 RA 酒店是 S 城唯一的酒店，那么 S 城的游客数量是不可能比 H 城数量多的。毕竟 RA 酒店的游客数量要少于 H 城最大酒店的游客数量。

(C) H 城有几家酒店比 RA 酒店大。无法确定 H 城其他酒店的情况。

(D) 一些在 H 城的酒店住过的游客也住过 RA 酒店。无法确定游客的重合度。

(E) H 城的一些酒店每年接待的游客比 RA 酒店少。无法确定 H 城其他酒店的情况。

5. Most cable television companies currently require customers to subscribe to packages of channels, but consumer groups have recently proposed legislation that would force the companies to offer a la carte pricing. Subscribers would pay less, argue the consumer groups, because they could purchase only the desired channels. However, the cable industry argues that under the current package pricing, popular channels subsidize less-popular ones, providing more options for viewers. For this reason, the industry claims that it is always cheaper for the consumer to purchase many bundled channels than to buy them individually.

Which of the following would be most important for the government to determine before deciding whether to require cable television companies to offer a la carte pricing in order to reduce consumer costs?

(A) Whether the total number of channels offered to consumers would decrease, along with programming diversity, as a result of the a la carte pricing structure

(B) Whether advertising revenue for the cable television companies would decrease as a result of the a la carte pricing structure

(C) Whether the vast majority of consumers would greatly reduce the number of channels purchased if given the option of purchasing them individually

(D) Whether cable and satellite companies currently have the ability to buy channels individually from programmers and content providers

(E) Whether a la carte subscribers would be required to have new television set-top boxes

类别：方案

目标：为了降低客户的成本。

方案：要求有线电视公司提供点菜式定价。

推理：答案选项需指出“有线电视公司是否可以提供点菜式定价”或“提供点菜式定价能否真的可以降低客户的成本”。

选项分析：

(A) 提供给消费者的频道总数是否会因为点菜式定价结构而减少，以及节目的多样性是否会减少节目的多样性与成本无关。

(B) 有线电视公司的广告收入是否会因为点菜式的定价结构而减少。与广告收入无关。

(C) 正确。如果可以选择单独购买频道，绝大多数消费者是否会大大减少购买频道的数量。如果可以单独购买频道，消费者还是会购买和之前一样多的频道，那肯定是打包买会更便宜；如果可以单独购买频道，消费者买的频道数量会大大减少，那消费者确实可以少花钱，不再需要为不想看的频道买单。

(D) 有线卫星公司目前是否有能力从节目主持人和内容提供商那里单独购买频道。与有线卫星公司无关。

(F) 点带用户是否需要新的电视机顶盒。即使需要买新的机顶盒，我们也不知道机顶盒的成本加上单独购买频道的成本是否会高于打包购买的成本。所以，此选项无法帮我们判断客户的成本究竟会不会降低。

6. New water-current patterns have brought more mixing of the Thalian Sea's shallow waters with nutrient-rich deeper waters. For coral and algae, both of which inhabit the shallow waters, growth rates depend principally on availability of nutrients and sufficient sunlight. The algae's rate of growth has increased, yet the coral's growth has slowed. Since algae can block sunlight from reaching coral below, the slowed coral growth is probably attributable to increased algae growth.

Which of the following, if true, most strengthens the argument?

(A) The new water-current patterns in the Thalian Sea have increased the concentrations of industrial pollutants in its shallower parts.

(B) In those areas of the Thalian Sea where there has been the greatest increase in algae growth, the growth of coral has stopped altogether.

(C) In some areas of the Thalian Sea where shallow water receives abundant sunlight, there is no coral and there are only small amounts of algae.

(D) Many of the nutrients essential to the growth of algae play no role in the growth of coral.

(E) The mixing of the sea's shallow waters with nutrient-rich deeper waters is expected to diminish after this year.

类别：因果推理

前提：藻类的生长速度加快了，但珊瑚的生长速度却减缓了。

结论：珊瑚的生长放缓可能是由于藻类的增长造成的（藻类增长导致了珊瑚生长放缓）。

推理：

纯粹巧合：在其他场景下，“藻类的增长”和“珊瑚的生长放缓”同时存在；或存在能解释“藻类的增长”导致“珊瑚的生长放缓”的原理。

他因导致结果：没有其他导致“珊瑚的生长放缓”的原因。

因果倒置：不是“珊瑚的生长放缓”导致“藻类的增长”。

选项分析：

（A）塔利安海的新水流模式增加了其较浅部分的工业污染物的浓度。

（B）正确。在塔利安海藻类生长增长最快的那些地区，珊瑚的生长已经完全停止。本选项强调了因果两件事是同时发生的，即藻类多的地方，珊瑚不生长了。

（C）在塔利安海的一些地区，浅水区阳光充足，没有珊瑚，只有少量的藻类。

（D）藻类生长所必需的许多营养物质对珊瑚的生长没有作用。

（E）该海域的浅水区与营养丰富的深水区的混合，今年以后预计会减少。

7. Healthy lungs produce a natural antibiotic that protects them from infection by routinely killing harmful bacteria on airway surfaces. People with cystic fibrosis, however, are unable to fight off such bacteria, even though their lungs produce normal amounts of the antibiotic. The fluid on airway surfaces in the lungs of people with cystic fibrosis has an abnormally high salt concentration; accordingly, scientists hypothesize that the high salt concentration is what makes the antibiotic ineffective.

Which of the following, if true, most strongly supports the scientists' hypothesis?

(A) When the salt concentration of the fluid on the airway surfaces of healthy people is raised artificially, the salt concentration soon returns to normal.

(B) A sample of the antibiotic was capable of killing bacteria in an environment with an unusually low concentration of salt.

(C) When lung tissue from people with cystic fibrosis is maintained in a solution with a normal salt concentration, the tissue can resist bacteria.

(D) Many lung infections can be treated by applying synthetic antibiotics to the airway surfaces.

(E) High salt concentrations have an antibiotic effect in many circumstances.

类别：因果推理

前提：囊性纤维化患者肺部气道表面液体的盐浓度异常高。

结论：高盐浓度使抗生素无效。

推理：

纯粹巧合：在其他场景下，“高盐浓度”和“抗生素无效”同时存在；或存在能解释“高盐浓度”导致“抗生素无效”的原理。

他因导致结果：不存在其他导致“抗生素无效”的原因。

因果倒置：不是“抗生素无效”导致“高盐浓度”。

选项分析：

(A) 当人为提高健康人气道表面液体的盐浓度的时候，盐浓度马上就会恢复至正常水平。本选项讨论的是正常人的气道不会具有高盐浓度的液体，与推理文段中的因果没有关系。

(B) 这种抗生素的样品可以在盐浓度很低的情况下杀死细菌。本选项描述的是低盐浓度时抗生素的表现，与高盐浓度无关。

(C) 正确。如果从囊性纤维化患者的身上切下来的肺组织被放到正常盐浓度的地方，那么该组织可以杀死病菌。通过控制变量，证明盐浓度和抗病能力的相关性，属于“不是纯粹巧合”。

(D) 许多肺部感染可以通过将合成抗生素用于气道表面来治愈。合成抗生素和自然抗生素没有关系。

(E) 高盐浓度在很多情况下具有抗生素的效果。本选项描述高盐浓度的一个属性，与讨论无关。

8. Brant Lumber employs numerous workers to saw logs into boards. Brant is considering refitting its sawmill with computerized log-cutting machines that would saw the logs automatically and that would increase the number of logs the sawmill could process per day. The refit would allow Brant to reduce its workforce, but it would not increase the sawmill's output, since the sawmill already processes as many logs per day as Brant's timber sources can supply.

Which of the following, if true, casts the most doubt on the analysis of the effects of the refit on the sawmill's output?

(A) The computerized log-cutting machines cost far more to purchase than the machines currently installed in the sawmill did when they were new.

(B) Replacing some of the workers at the sawmill with computerized log-cutting machines would enable Brant to increase profits, even though those machines would not be used to capacity.

(C) By relocating the sawmill instead of refitting it, Brant would be able to increase the supply of lops available for the sawmill to process into boards.

(D) Less of a log is wasted when it is cut into boards by computerized log-cutting machines than when it is cut into boards by workers using the sawmill's current equipment.

(E) The computerized log-cutting machines can be adjusted far more quickly to cut boards of different dimensions than can the sawmill's current equipment.

类别：普通预测推理

前提：锯木厂每天加工的原木数量已经达到布兰特木材来源所能提供的数量。

结论：这个改装不会增加锯木厂的产量。

推理：结论“这个改装不会增加锯木厂的产量”的充分条件应为“原材料不变的情况下转化率也不变，即废弃材料数量也不变”。

选项分析：

(A) 计算机化的原木切割机器的购买成本远远高于目前安装在锯木厂的机器在新买的时候的购买成本。与机器的购买成本无关。
(B) 用计算机化的原木切割机器取代锯木厂的一些工人，可以使布兰特增加利润，尽管这些机器不会被用得很好。与利润无关。
(C) 通过迁移锯木厂而不是对其进行改装，布兰特将能够增加可供锯木厂加工成板材的木片供应。与其他方法无关，现在只讨论"refit"这个方法。
(D) 正确。与工人使用锯木厂现有的设备将原木切割成木板相比，用计算机化的原木切割机将原木切割成木板时，浪费的木材更少。本选项直接指出了废弃木材会减少，那产量是有可能增加的。
(E) 与锯木厂现有的设备相比，计算机化的原木切割机可以更快地调整，以切割不同尺寸的木板。机器的工作情况与讨论无关。

9. Consultant: Innovation is fostered by frequent exchange and discussion of relevant ideas, but business work spaces divided into individual offices and cubicles encourage employees to work in relative isolation. So, employees whose work requires them to innovate will generally be more productive if assigned to work at desks in large office space without dividing barriers.

Which of the following is an assumption on which the consultant's argument depends?

(A) In larger office spaces without dividing barriers, employees whose work requires them to innovate would more often discuss relevant ideas with each other than in other office settings.
(B) Discussing ideas at work reduces the time that employees would otherwise spend alone working on innovative ideas.
(C) In work spaces divided into individual offices and cubicles, employees whose work does not require them to innovate are sometimes more productive than in large open work spaces.

(D) Most of the ideas employees exchange and discuss at work are consciously considered by those employees to be directly relevant to innovation in their work.

(E) Business consultants generally recommend that innovative employees be assigned to work at desks in larger office spaces without dividing barriers if that will make them more productive.

类别：普通预测推理

前提：需要创新的员工被安排在没有隔断的大型办公空间的办公桌上工作。

结论：他们的工作效率通常会更高。

推理：背景信息交代“创新需要频繁地交流和讨论相关想法”，所以结论“他们的工作效率通常会更高”的充分条件应为“他们会更频繁地交流和讨论相关想法”。答案选项要确保“员工被安排在没有隔断的大型办公空间的办公桌上工作”可以“更频繁地交流和讨论相关想法。”

选项分析：

(A) 正确。在没有隔断的大办公空间里，工作需要创新的员工会比在其他办公环境中更频繁地相互讨论相关想法。本选项直接修正了偏差。

(B) 在工作中讨论想法可以减少员工单独研究创新想法的时间。

(C) 在分为单独办公室和隔间的工作空间里，工作不需要创新的员工有时会比在大型开放式工作空间里效率更高。

(D) 员工在工作中交流和讨论的大多数想法都被这些员工有意识地认为与他们工作中的创新直接相关。

(E) 商业顾问一般建议，将创新型员工分配到没有隔断障碍的更大的办公空间的办公桌上工作，这样会使他们的工作效率更高。

10. In January of last year the Moviemania chain of movie theaters started propping its popcorn in canola oil, instead of the less healthful coconut oil that it had been using until then. Now Moviemania is planning to switch back, saying that the change has hurt popcorn sales. That claim is false, however, since according to Moviemania's own sales figures, Moviemania sold 5 percent more popcorn last year than in the previous year.

Which of the following, if true, most strongly supports the argument against Moviemania's claim?

(A) Total sales of all refreshments at Moviemania's movie theaters increased by less than 5 percent last year.

(B) Moviemania makes more money on food and beverages sold at its theaters than it does on sales of movie tickets.

(C) Moviemania's customers prefer the taste of popcorn popped in coconut oil to that of popcorn popped in canola oil.

(D) Total attendance at Moviemania's movie theaters was more than 20 percent higher last year than the year before.

(E) The year before last, Moviemania experienced a 10 percent increase in popcorn sales over the previous year.

类别：普通归因推理

前提：Moviemania 去年的爆米花销量比前年多 5%。

结论：改变没有损害爆米花的销售。

推理："改变没有损害爆米花的销售"的其他证据存在，或给出其他能解释"爆米花销量比前年多 5%"的原因不存在。

选项分析：

（A）正确。去年 Moviemania 电影院所有茶点的总销量增加了不到5%。如果所有东西的销量普遍涨很多，那么就可能是因为整体的市场好了，所以爆米花的销量才提高。但现在其他东西的销量涨幅不如爆米花，说明爆米花是因为自身的原因销量在上涨，改变确实没有损害爆米花的销售。

（B）Moviemania 在其影院销售的食品和饮料上赚的钱比销售电影票赚的钱多。

（C）与菜籽油爆米花相比，Moviemania 的顾客更喜欢椰子油爆米花的味道。

（D）去年 Moviemania 电影院的总上座率比前年高出 20% 以上。

（E）前年，Moviemania 的爆米花销售额比前一年增加了 10%。

第三组

1. We know very little about the Etruscans, whose civilization flourished in central Italy from the ninth to the second centuries BC, and much of what we do know comes from their art, for the Etruscan language is all but ciphers to us. Greek and Phoenician sources indicate that the Etruscans enjoyed trade and commerce with those societies, by which the Etruscans were doubtless influenced. Yet unlike women in those societies, women in Etruscan society participated in important social functions, as documented by several Etruscan artworks. Thus, the Etruscan women held unusual status and freedom for the time and region.

The argument above requires which of the following assumptions?

(A) The Etruscan artists who created the artworks in question did not intend them as depictions of fantasy scenarios.

(B) Some of the Etruscan artworks in question have been found in the remains of Greek and Phoenician cities.

(C) An Etruscan's ability to participate in an important social function was inversely related to the level of that person's status and freedom.

(D) Women's place in Etruscan society would be clearer to us if we could understand the Etruscan language.

(E) Greek and Phoenician artworks depicting important functions sometimes show women participating.

2. Aroca City currently funds its public schools through taxes on property. In place of this system, the city plans to introduce a sales tax of three percent on all retail sales in the city. Critics protest that **three percent of current retail sales**

falls short of the amount raised for schools by property taxes. The critics are correct on this point. Nevertheless, implementing the plan will probably not reduce the money going to Aroca's schools. Several large retailers have selected Aroca City as the site for huge new stores, and these are certain to draw large numbers of shoppers from neighboring municipalities, where sales are taxed at rates of six percent and more. In consequence, **retail sales in Aroca City are bound to increase substantially**.

In the argument given, the two potions in boldface play which of the following roles?

(A) The first is an objection that has been raised against a certain plan; the second is a prediction that, if accurate, undermines the force of that objection.

(B) The first is a criticism, endorsed by the argument, of a funding plan; the second is a point the argument makes in favor of adopting a alternative plan.

(C) The first is a criticism, endorsed by the argument, of a funding plan; the second is the main reason cited by the argument for its endorsement of the criticism.

(D) The first is a claim that the argument seeks to refute; the second is the main point used by the argument to show that the claim is false.

(E) The first is a claim that the argument accepts with certain reservations; the second presents that claim in a rewarding that is not subject to those reservations.

3. Which of the following, if true, would most logically complete the argument?

Researchers studying the Greenland ice shelf were shocked to discover the presence of water, colored pink, two miles under the ice. They concluded that the pink color was probably due to microorganisms. They also concluded that the water was likely due to a drastic increase in global warming, an increase that has been occurring for just a few hundred years. But these conclusions may not be correct. It is just as possible that the water has accumulated under the ice for thousands of years, since the ice could have acted as a perfect insulator to trap heat emanating from the earth. Moreover, ________.

(A) it would take much longer than a few hundred years for such an ecosystem of microorganisms to establish itself

(B) recent evidence suggests that the rate of global warming is increasing rapidly as a result of human activities

(C) some scientists believe that the increase in global warming in the last few hundred years has been even greater than is generally acknowledged

(D) similar areas of pink water have been found under ice shelves in Antarctica

(E) scientists have in recent years discovered microorganisms in locations where they had believed no life could survive

4. Two species of coffee are commercially grown: arabica, the original source of the drink, and robusta, which has more recently come under cultivation. The proportion of robusta beans in inexpensive blended coffees has increased in recent decades because robusta resists frost and disease better, fruits faster, and grows at lower elevations. Expensive gourmet coffees typically contain only arabica beans because robusta beans, though higher in caffeine than arabicas, are more neutral in flavor and consequently less interesting.

The information given most strongly supports which of the following?

(A) Robusta cannot be grown successfully in the regions where arabica is commercially cultivated.

(B) Inexpensive coffees sold now contain more caffeine than did coffees sold a few decades ago.

(C) The only factor determining differences in flavor among different coffees is the proportion of robusta beans that they contain.

(D) Although more robusta than arabica beans are sold, the total value of arabica beans sold each year exceeds that of robusta beans.

(E) Arabica and robusta are the only species whose beans can be used to make coffee.

5. To reduce the danger to life and property posed by major earthquakes, scientists have been investigating several techniques for giving advance warning of dangerous earthquakes. Since catfish swim erratically before earthquakes, some investigators have proposed monitoring catfish to predict dangerous earthquakes.

Which of the following, if true, most seriously undermines the usefulness of the proposal?

(A) In Japan, which is subject to frequent earthquakes, the behavior of catfish has long been associated with earthquakes.

(B) Mechanical methods for detecting earthquakes have not proved effective.

(C) Tremors lead to the release of hydrogen sulfide gas into water, thereby causing various fish and shellfish to behave erratically.

(D) Careful construction can reduce the dangers posed by earthquakes.

(E) Even very slight, fleeting tremors cause catfish to swim erratically.

6. When feeding, aquatic birds known as phalaropes often spin rapidly on the water's surface, pecking for food during each revolution. To execute these spins, phalaropes kick one leg harder than the other. This action creates upwelling currents. Because plankton on which phalaropes feed typically occurs in greater quantities well below the surface, it is hypothesized that by

spinning phalaropes gain access to food that would otherwise be beyond their reach.

Which of the following, if true, most strongly supports the hypothesis?

(A) Phalaropes rarely feed while on land.
(B) A given phalarope spins exclusively either to the right or to the left.
(C) Phalaropes sometimes spin when they are not feeding.
(D) Different phalaropes spin at somewhat different rates.
(E) Phalaropes do not usually spin when food is abundant at the surface.

7. Pretzel vendor: The new license fee for operating a pretzel stand outside the art museum is prohibitively expensive. Charging typical prices, a vendor would need to sell an average of 25 pretzels per hour to break even. At my stand outside the city hall, I average only 15 per hour. Therefore, I could not break even running a pretzel stand outside the art museum, much less turn a profit.

Which of the following, if true, most strongly supports the pretzel vendor's argument?

(A) There is currently no license fee for operating a pretzel stand outside the city hall.
(B) Pretzel vendors who operate stands outside the art museum were making a profit before the imposition of the new license fee.
(C) The number of pretzel stands outside the art museum is no greater than the number of pretzel stands now outside the city hall.
(D) People who buy pretzels at pretzel stands are most likely to do so during the hours at which the art museum is open to the visitors.
(E) Fewer people passing the art museum than passing city hall are to buy pretzels.

8. Sea turtles caught in traditional shrimp nets drown. Five years ago, Ridland's government introduced a requirement that all nets used to catch shrimp be equipped with special devices that allow sea turtles to escape if caught in the net. In the first four years, significantly fewer sea turtles washed up dead on Ridland's shores. Last year, the number was five percent higher than before the program began. Clearly, therefore, many shrimpers have stopped complying with the law.

Which of the following, if true, most seriously weakens the argument?

(A) Because of conservation programs at sea turtle nesting sites, the population of sea turtles migrating through Ridland's waters was up sharply last year.

(B) The special devices allow not only turtles but also some shrimp to escape from shrimp nets, reducing the amount of shrimp that a shrimp boat can catch in a day.

(C) Because of the direction of ocean currents in and around Ridland's waters, any sea turtles that wash up dead on Ridland's shores are likely to have died within the area fished by Ridlandian shrimp boats.

(D) The fine for violating the law requiring shrimp nets to have the special devices was increased last year and is now greater than the cost of purchasing such a device.

(E) Because of a number of measures unrelated to the protection of sea turtles, Ridland's waters are less polluted now than they were five years ago.

9. Which of the following most logically completes the passage?

Commentator: The number of people who report that they have been seriously depressed at least once in their lives is three times as great today as it was 50 years ago. This statistic is often used to support the view that people no longer believe in the validity of certain social institutions that once gave individuals a sense of psychological stability. How much support the statistic provides is an open question, however, since _______.

(A) many social institutions are not considered important in promoting individuals' sense of psychological stability

(B) increased public awareness about mental health issues has eliminated the stigma that was associated with admitting to feelings of depression 50 years ago

(C) the number of people who seek treatment for serious depression is greater now than it was 50 years ago

(D) many of the social institutions that once gave individuals a sense of psychological stability still exist as social institutions

(E) the clinical definition of what constitutes serious depression has remained unchanged over the course of the past 50 years

10. Because of its increasing reliance on computer networks, MegaCorp faces an increasing security threat from illegal infiltration of its computer systems, by its own employees and by outsiders. Accordingly, MegaCorp last year implemented sophisticated and expensive new safeguards to prevent such security breaches. The company reported substantially fewer instances of security breaches to the police this year than last year, so evidently the safeguards have been highly effective.

Which of the following, if true, most seriously weakens the argument?

(A) It is impossible for a business as large as MegaCorp to safeguard its computer systems against all potential security breaches.

(B) Most of the computer-system security breaches reported to police by MegaCorp last year were committed by outsiders.

(C) The cost of implementing computer safeguards is generally much lower than the potential cost of computer security breaches.

(D) Spending large sums of money on new safeguards often makes the leaders of a company reluctant to report breaches of those safeguards.

(E) MegaCorp began making extensive use of computer networks much later than most of its competitors.

答案及解析

1. We know very little about the Etruscans, whose civilization flourished in central Italy from the ninth to the second centuries BC, and much of what we do know comes from their art, for the Etruscan language is all but ciphers to us. Greek and Phoenician sources indicate that the Etruscans enjoyed trade and commerce with those societies, by which the Etruscans were doubtless influenced. Yet unlike women in those societies, women in Etruscan society participated in important social functions, as documented by several Etruscan artworks. Thus, the Etruscan women held unusual status and freedom for the time and region.

The argument above requires which of the following assumptions?

(A) The Etruscan artists who created the artworks in question did not intend them as depictions of fantasy scenarios.

(B) Some of the Etruscan artworks in question have been found in the remains of Greek and Phoenician cities.

(C) An Etruscan's ability to participate in an important social function was inversely related to the level of that person's status and freedom.

(D) Women's place in Etruscan society would be clearer to us if we could understand the Etruscan language.

(E) Greek and Phoenician artworks depicting important functions sometimes show women participating.

类别：泛化推理

前提：据一些艺术作品记录，E 国的女性参加了重要的社会活动。

结论：E 国的女性拥有不寻常的社会地位和自由。

推理：指出艺术作品上的记录可以代表整个社会的实际情况，或者除了艺术作品上的记录，是否有其他记录可以证明 E 国的女性拥有不寻常的社会地位和自由。

选项分析：

(A) 正确。创作艺术品的艺术家没有故意描绘一个美好的场景。此选项直接指出并不是艺术家故意这样做的，艺术作品的情况确实可以代表实际情况。

(B) 一些艺术品在 G 和 P 的一些城市也发现了。艺术作品在何处发现与讨论无关。

(C) E 国人参与重要社会活动的能力与这个人的地位、自由是负相关的。如果参与重要社会活动意味着其社会地位低，那么结论就不成立了。本选项直接削弱了结论。

(D) 如果我们能理解 E 国语言，那么 E 国女性的地位会更清晰。女性地位在什么条件下会更清晰，与艺术作品能否代表实际情况无关。

(E) 描绘重要功能的 G 和 P 的艺术品有时展示出女性的参与。艺术作品本身的内容与讨论无关。

2. Aroca City currently funds its public schools through taxes on property. In place of this system, the city plans to introduce a sales tax of three percent on all retail sales in the city. Critics protest that **three percent of current retail sales falls short of the amount raised for schools by property taxes**. The critics are correct on this point. Nevertheless, implementing the plan will probably not reduce the money going to Aroca's schools. Several large retailers have selected Aroca City as the site for huge new stores, and these are certain to draw large numbers of shoppers from neighboring municipalities, where sales are taxed at rates of six percent and more. In consequence, **retail sales in Aroca City are bound to increase substantially**.

In the argument given, the two potions in boldface play which of the following roles?

(A) The first is an objection that has been raised against a certain plan; the second is a prediction that, if accurate, undermines the force of that objection.

(B) The first is a criticism, endorsed by the argument, of a funding plan; the second is a point the argument makes in favor of adopting a alternative plan.

(C) The first is a criticism, endorsed by the argument, of a funding plan; the second is the main reason cited by the argument for its endorsement of the criticism.

(D) The first is a claim that the argument seeks to refute; the second is the main point used by the argument to show that the claim is false.

(E) The first is a claim that the argument accepts with certain reservations; the second presents that claim in a rewarding that is not subject to those reservations.

类别：分析论证

推理：阅读文段，可以确定主结论为：Implementing the plan will probably not reduce the money going to Aroca's schools。第一个黑体句是主结论同意的事实，第二个黑体句是支持主结论的前提。

选项分析：

(A) 正确。第一个黑体句是对某一方案提出的反对意见；第二个黑体句是一个预测，如果准确的话，会削弱该反对意见。

(B) 第一个黑体句是对某项筹资方案的批评，并得到了论证的支持；第二个黑体句是论证支持采用替代方案的观点。文段并没有“支持”第一个黑体句中的批评，只是承认消费税率确实比房产税率低这一事实。

(C) 第一个黑体句是对某项筹资方案的批评，并得到了论证的支持；第二个黑体句是论证为认可该批评所引用的主要理由。

(D) 第一个黑体句是论证试图反驳的主张；第二个黑体句是论证用来证明该主张是错误的主要观点。本选项主要错在论证并没有反对第一个黑体句，而是同意这个黑体句的说法。

(E) 第一个黑体句是论证接受的主张，但有一定的保留意见；第二个黑体句是以不受这些保留意见影响的奖励方式提出该主张。第二个黑体句并不是第一个黑体句的呈现方式。

3. Which of the following, if true, would most logically complete the argument?

Researchers studying the Greenland ice shelf were shocked to discover the presence of water, colored pink, two miles under the ice. They concluded that the pink color was probably due to microorganisms. They also concluded that the water was likely due to a drastic increase in global warming, an increase that has been occurring for just a few hundred years. But these conclusions may not be correct. It is just as possible that the water has accumulated under the ice for thousands of years, since the ice could have acted as a perfect insulator to trap heat emanating from the earth. Moreover, ________.

(A) it would take much longer than a few hundred years for such an ecosystem of microorganisms to establish itself
(B) recent evidence suggests that the rate of global warming is increasing rapidly as a result of human activities
(C) some scientists believe that the increase in global warming in the last few hundred years has been even greater than is generally acknowledged
(D) similar areas of pink water have been found under ice shelves in Antarctica
(E) scientists have in recent years discovered microorganisms in locations where they had believed no life could survive

类别：构建论证中的“现象解释”

推理：阅读文段，文中需要解释的矛盾点为：粉色并不是由微生物造成的，以及这些水也不是由全球变暖加剧造成的。

选项分析：

(A) 正确。这样一个微生物生态系统需要比几百年更长的时间才能建立起来。

(B) 最近的证据表明，由于人类活动的影响，全球变暖的速度正在迅速增加。

(C) 一些科学家认为，在过去的几百年里，全球变暖的加剧程度比人们普遍认为的还要大。

(D) 在南极洲的冰架下也发现了类似的粉红色水域。

(E) 近年来，科学家们在他们认为没有生命可以生存的地方发现了微生物。

4. Two species of coffee are commercially grown: arabica, the original source of the drink, and robusta, which has more recently come under cultivation. The proportion of robusta beans in inexpensive blended coffees has increased in recent decades because robusta resists frost and disease better, fruits faster, and grows at lower elevations. Expensive gourmet coffees typically contain only arabica beans because robusta beans, though higher in caffeine than arabicas, are more neutral in flavor and consequently less interesting.

The information given most strongly supports which of the following?

(A) Robusta cannot be grown successfully in the regions where arabica is commercially cultivated.

(B) Inexpensive coffees sold now contain more caffeine than did coffees sold a few decades ago.

(C) The only factor determining differences in flavor among different coffees is the proportion of robusta beans that they contain.

(D) Although more robusta than arabica beans are sold, the total value of arabica beans sold each year exceeds that of robusta beans.

(E) Arabica and robusta are the only species whose beans can be used to make coffee.

类别：构建论证中的“确定结论”

推理：

阅读原文，文段大意为：两种咖啡豆：A 和 R。

A 的特点是：咖啡饮料的原始来源，昂贵的咖啡只含有 A。

R 的特点是：最近才兴起，在廉价混合咖啡中的比例增加，能更好地抵抗霜冻和疾病，结果实更快，而且生长在低海拔地区，咖啡因含量更高，但味道更中性，没那么有意思。

选项分析：

(A) R 不能在 A 商业化种植的地区成功种植。原文没提及。

(B) 正确。现在出售的廉价咖啡比几十年前出售的咖啡含有更多的咖啡因。结合“R 在廉价咖啡中的比例增加”和“R 的咖啡因含量更高”可以得出此选项。

(C) 决定不同咖啡风味差异的唯一因素是它们所含的 R 的比例。原文没提咖啡风味的唯一决定性因素。

(D) 虽然销售的 R 比 A 多，但每年销售的 A 的总价值超过了 R。原文没提销售总价值。

(E) A 和 R 是唯一可以用来制作咖啡的品种。原文没提及。

5. To reduce the danger to life and property posed by major earthquakes, scientists have been investigating several techniques for giving advance warning of dangerous earthquakes. Since catfish swim erratically before earthquakes, some investigators have proposed monitoring catfish to predict dangerous earthquakes.

Which of the following, if true, most seriously undermines the usefulness of the proposal?

(A) In Japan, which is subject to frequent earthquakes, the behavior of catfish has long been associated with earthquakes.

(B) Mechanical methods for detecting earthquakes have not proved effective.

(C) Tremors lead to the release of hydrogen sulfide gas into water, thereby causing various fish and shellfish to behave erratically.

(D) Careful construction can reduce the dangers posed by earthquakes.

(E) Even very slight, fleeting tremors cause catfish to swim erratically.

类别：方案

目标：预测危险的地震。

方案：监控鲶鱼。

推理：答案选项需指出“无法监控鲶鱼”或“监控鲶鱼无法预测危险的地震”。

选项分析：

(A) 在地震频发的日本，鲶鱼的行为一直与地震联系在一起。此选项加强了鲶鱼和地震的关系。

(B) 用于探测地震的机械方法尚未证明是有效的。与机械方法无关。

(C) 震颤导致硫化氢气体释放到水中，从而导致各种鱼类和贝类的行为不稳定。此选项给出了方案达到目标的原理，一定程度上可以加强推理。

(D) 精心建造的建筑可以减少地震带来的危险。与建筑无关。

(E) 正确。即使是非常轻微的、短暂的震颤也会导致鲶鱼不规律地游动。指出监控鲶鱼可能无法达到目的。

6. When feeding, aquatic birds known as phalaropes often spin rapidly on the water's surface, pecking for food during each revolution. To execute these spins, phalaropes kick one leg harder than the other. This action creates upwelling currents. Because plankton on which phalaropes feed typically occurs in greater quantities well below the surface, it is hypothesized that by spinning phalaropes gain access to food that would otherwise be beyond their reach.

Which of the following, if true, most strongly supports the hypothesis?

(A) Phalaropes rarely feed while on land.

(B) A given phalarope spins exclusively either to the right or to the left.

(C) Phalaropes sometimes spin when they are not feeding.

(D) Different phalaropes spin at somewhat different rates.

(E) Phalaropes do not usually spin when food is abundant at the surface.

类别：因果推理

前提：在进食时，被称为 phalarope 的水鸟经常在水面上快速旋转，在每次旋转中啄食。

结论：通过旋转，phalarope 获得了原本无法获得的食物。

推理：

纯粹巧合：在其他场景下，“旋转”和“phalarope 获得食物”同时存在；或存在“phalarope 获得食物”导致“旋转”的原理。
他因导致结果：没有其他导致“旋转”的原因。
因果倒置：不是“旋转”导致“phalarope 获得食物”。

选项分析：

(A) Phalaropes 很少在陆地上进食。陆地上的食物显然和水下的食物不同。

(B) Phalaropes 可能会向左转，也可能会向右转。向哪里转和推理文段的因果无关。

(C) Phalaropes 有时不进食的时候也旋转。本选项属于削弱，但题目问的是加强。如果鸟无论吃不吃饭都会转圈，那么鸟的转圈就和吃饭之间是纯粹巧合，并非具备因果关系。

(D) 不同的 phalaropes 旋转的速度不同。旋转速度和推理文段的因果无关。

(E) 正确。当水面上的食物比较丰富时，phalaropes 很少转圈。本选项是加强。当吃饭不需要转圈的时候鸟就不转圈，这表明了“吃饭”和“转圈”不是纯粹巧合的。

7. Pretzel vendor: The new license fee for operating a pretzel stand outside the art museum is prohibitively expensive. Charging typical prices, a vendor would need to sell an average of 25 pretzels per hour to break even. At my stand outside the city hall, I average only 15 per hour. Therefore, I could not break even running a pretzel stand outside the art museum, much less turn a profit.

Which of the following, if true, most strongly supports the pretzel vendor's argument?

(A) There is currently no license fee for operating a pretzel stand outside the city hall.

(B) Pretzel vendors who operate stands outside the art museum were making a profit before the imposition of the new license fee.

(C) The number of pretzel stands outside the art museum is no greater than the number of pretzel stands now outside the city hall.

(D) People who buy pretzels at pretzel stands are most likely to do so during the hours at which the art museum is open to the visitors.

(E) Fewer people passing the art museum than passing city hall are to buy pretzels.

类别：类比推理

前提：一个小贩平均每小时需要卖出 25 个椒盐卷饼才能达到收支平衡。在市政厅外的摊位上，我平均每小时只卖出 15 个椒盐卷饼。

结论：我在艺术博物馆外经营一个椒盐卷饼摊无法实现收支平衡，更不用说盈利了。

推理：考虑市政厅和艺术博物馆的其他相关相似点不缺失。

选项分析：

(A) 目前，在市政厅外经营一个椒盐卷饼摊是不收取执照费的。市政厅外是否收取执照费与在艺术博物馆外卖椒盐卷饼能否回本无关。

(B) 在征收新的许可证费用之前，在艺术博物馆外经营摊位的椒盐卷饼摊贩已经盈利了。与征收许可费之前的情况无关。

(C) 艺术博物馆外面椒盐卷饼摊的数量并不比现在市政厅外面椒盐卷饼摊的数量多。与摊贩数量无关。

(D) 在椒盐卷饼摊购买椒盐卷饼的人最有可能在艺术博物馆对游客开放的时间购买。与游客什么时间购买椒盐卷饼无关。

(E) 正确。经过艺术博物馆买椒盐卷饼的人比经过市政厅买椒盐卷饼的人少。买的人少更不可能挣到钱了。

8. Sea turtles caught in traditional shrimp nets drown. Five years ago, Ridland's government introduced a requirement that all nets used to catch shrimp be equipped with special devices that allow sea turtles to escape if caught in the net. In the first four years, significantly fewer sea turtles washed up dead on Ridland's shores. Last year, the number was five percent higher than before the program began. Clearly, therefore, many shrimpers have stopped complying with the law.

Which of the following, if true, most seriously weakens the argument?

(A) Because of conservation programs at sea turtle nesting sites, the population of sea turtles migrating through Ridland's waters was up sharply last year.

(B) The special devices allow not only turtles but also some shrimp to escape from shrimp nets, reducing the amount of shrimp that a shrimp boat can catch in a day.

(C) Because of the direction of ocean currents in and around Ridland's waters, any sea turtles that wash up dead on Ridland's shores are likely to have died within the area fished by Ridlandian shrimp boats.

(D) The fine for violating the law requiring shrimp nets to have the special devices was increased last year and is now greater than the cost of purchasing such a device.

(E) Because of a number of measures unrelated to the protection of sea turtles, Ridland's waters are less polluted now than they were five years ago.

类别：归因推理

前提：在最初的四年里，被冲到里德兰海岸的海龟明显减少。去年，这个数字比该方案开始前高出5%。

结论：许多捕虾者已经不再遵守该法律。

推理："许多捕虾者已经不再遵守该法律"的其他证据不存在，或给出其他能解释"这个数字比该方案开始前高出5%"的原因。

选项分析：

(A) 正确。由于海龟筑巢地的保护方案，去年通过里德兰水域迁徙的海龟数量急剧上升。本选项表明不一定是捕虾者不遵守法律，而可能是海龟的基数比以前更多了，从而导致去年的数字上升。

(B) 这些特殊的装置不仅让海龟，也让一些虾从捕虾网中逃脱，减少了捕虾船在一天内能捕到的虾的数量。

(C) 由于里德兰水域及其周围的洋流方向，任何被冲到里德兰海岸的海龟都可能是在里德兰捕虾船捕捞的区域内死亡的。

(D) 违反要求捕虾网必须装有特殊装置的法律的罚款去年有所增加，现在已经超过了购买这种装置的费用。

(E) 由于采取了一些与保护海龟无关的措施，现在里德兰水域比五年前污染要少。

9. Which of the following most logically completes the passage?

Commentator: The number of people who report that they have been seriously depressed at least once in their lives is three times as great today as it was 50 years ago. This statistic is often used to support the view that people no longer believe in the validity of certain social institutions that once gave individuals a sense of psychological stability. How much support the statistic provides is an open question, however, since ________.

(A) many social institutions are not considered important in promoting individuals' sense of psychological stability

(B) increased public awareness about mental health issues has eliminated the stigma that was associated with admitting to feelings of depression 50 years ago

(C) the number of people who seek treatment for serious depression is greater now than it was 50 years ago

(D) many of the social institutions that once gave individuals a sense of psychological stability still exist as social institutions

(E) the clinical definition of what constitutes serious depression has remained unchanged over the course of the past 50 years

类别：构建论证中的“现象解释”

推理：

阅读文段，需要解释的矛盾点为：今天，报告说自己一生中至少有一次严重抑郁的人数是50年前的三倍，但这个现象的背后原因并不是曾经给个人带来心理稳定感的某些社会机构不再有效了。

给出一个他因，能合理解释“一生中至少有一次严重抑郁的人数是50年前的三倍”即可。

选项分析：

(A) 许多社会机构在促进个人的心理稳定方面被认为不重要。机构的重要性与讨论无关。

(B) 正确。公众对心理健康问题的认识提高，消除了50年前承认抑郁的耻辱感。本选项可以解释为什么报告自己抑郁的人更多，且不是因为社会机构无效。

(C) 现在寻求治疗严重抑郁的人数比 50 年前更多了。寻求治疗的人数与讨论无关。

(D) 许多曾经给予个人心理稳定感的社会机构，现在仍然作为社会机构存在着。与这些机构无关。

(E) 在过去的 50 年里，关于什么是严重抑郁的临床定义一直没有变化。如果定义变化，可能是导致人数变多的他因。

10. Because of its increasing reliance on computer networks, MegaCorp faces an increasing security threat from illegal infiltration of its computer systems, by its own employees and by outsiders. Accordingly, MegaCorp last year implemented sophisticated and expensive new safeguards to prevent such security breaches. The company reported substantially fewer instances of security breaches to the police this year than last year, so evidently the safeguards have been highly effective.

Which of the following, if true, most seriously weakens the argument?

(A) It is impossible for a business as large as MegaCorp to safeguard its computer systems against all potential security breaches.

(B) Most of the computer-system security breaches reported to police by MegaCorp last year were committed by outsiders.

(C) The cost of implementing computer safeguards is generally much lower than the potential cost of computer security breaches.

(D) Spending large sums of money on new safeguards often makes the leaders of a company reluctant to report breaches of those safeguards.

(E) MegaCorp began making extensive use of computer networks much later than most of its competitors.

类别：归因推理

前提：该公司今年向警方报告的安全漏洞事件比去年少得多。

结论：这些保障措施是非常有效的。

推理："这些保障措施有效"的其他证据不存在，或有其他能够解释"该公司今年向警方报告的安全漏洞事件比去年少得多"的原因。

选项分析：

(A) 像 MegaCorp 这样大的企业不可能保护其计算机系统不受所有潜在安全漏洞的影响。

(B) 去年，MegaCorp 向警方报告的大多数计算机系统安全漏洞都是由外部人员实施的。

(C) 实施计算机保障措施的成本通常比计算机安全漏洞的潜在成本低得多。

(D) 正确。在新的保障措施上花费大量资金，往往使公司的领导人不愿意报告这些保障措施存在漏洞的情况。本选项给出了一个报告安全漏洞事件少的另一个可能的解释，即可能是大家不愿意报告，而非这些保护措施有效。

(E) MegaCorp 开始广泛使用计算机网络的时间比其大多数竞争对手晚得多。